# Keeping in Time With Your Body Clock

# Keeping in Time With Your Body Clock

## J. M. Waterhouse
Senior Lecturer, Research Institute for Sport and
Exercise Sciences, Liverpool John Moores University

## D. S. Minors
Senior Lecturer, Department of Biological
Sciences, University of Manchester

## M. E. Waterhouse
Lecturer in English, City College, Manchester

## T. Reilly
Director, Research Institute for Sport and Exercise
Sciences, Liverpool John Moores University

## G. Atkinson
Senior Lecturer, School for Health,
University of Durham

OXFORD
UNIVERSITY PRESS

*This book has been printed digitally and produced in a standard specification
in order to ensure its continuing availability*

# OXFORD
UNIVERSITY PRESS

Great Clarendon Street, Oxford OX2 6DP

Oxford University Press is a department of the University of Oxford.
It furthers the University's objective of excellence in research, scholarship,
and education by publishing worldwide in

Oxford New York

Auckland   Cape Town   Dar es Salaam   Hong Kong   Karachi
Kuala Lumpur   Madrid   Melbourne   Mexico City   Nairobi
New Delhi   Shanghai   Taipei   Toronto
With offices in
Argentina   Austria   Brazil   Chile   Czech Republic   France   Greece
Guatemala   Hungary   Italy   Japan   South Korea   Poland   Portugal
Singapore   Switzerland   Thailand   Turkey   Ukraine   Vietnam

Oxford is a registered trade mark of Oxford University Press
in the UK and in certain other countries

Published in the United States
by Oxford University Press Inc., New York

© Oxford University Press 2002

The moral rights of the author have been asserted

Database right Oxford University Press (maker)

Reprinted 2009

ISBN 978-0-19-851074-1

# Preface

Hidden deep within our brain we have a body clock. Its activities help to determine when we wake up in the morning and when we go to sleep at night. If we are awake, it increases our ability to perform mental and physical tasks, it alters our body temperature and how cold we feel, and it influences our digestive system and how healthy our appetite is. It also affects our hormones, heart and blood pressure, and even those times when we are most likely to have an asthma attack.

Our body clock is responsible for enabling us to fit into an environment that is dominated by the 24-hour rhythm produced by sunlight falling upon our rotating Earth.

But the body clock can go wrong or be tampered with by our modern life-style—our body clock is important in jet-lag, the malaise of shift-work, and in some forms of insomnia, and so can affect any of us.

Recent scientific research has enabled us to piece together how the body clock works and how it is controlled in healthy people. As a result, we now have a better understanding of certain illnesses and subsequently can improve the treatment of some clinical disorders. Also, we can offer advice to travellers across time zones and to night-shift workers.

*Keeping in Time With Your Body Clock* is an introductory and yet up-to-date account of this rapidly developing field of research. As everybody is influenced by their body clock, we suggest simple ways of testing these influences, as well as give advice that helps us to keep in time with our body clock.

JM                                                                June 2002

# Contents

# How to use this book

The book is divided into two main parts. In Part I, we consider the basic evidence which describes the presence and make-up of our body clock and the rhythms it produces. In Part II, we apply this knowledge to an understanding of:

- problems that can arise when the system goes wrong;
- difficulties that follow after we have changed our lifestyle, such as after a time-zone transition or during shift work;
- the usefulness of rhythms in clinical practice.

In addition, we want the reader to become as involved in the field as possible and to gain some idea of how the scientific method works. Therefore, at several points in the text, we invite readers to test themselves, to try to solve some problems, and to make use of their knowledge of their own body clock.

# Part I
# Your body clock in health

Our subject is the body clock — how it influences our physiology and behaviour and how it interacts with the rhythms in our environment.

In this first part of the book we will attempt to answer the following questions:

- What scientific evidence leads us to believe that we possess an internal body clock?

- What do we know about the rhythms that the clock controls?

- Is the clock the same in all of us and at all stages of our lives?

- What do we know about the clock itself — where is it, what influences it, and how does it work?

- What is the usefulness of such a clock?

# Chapter 1

# The internal and external causes of body rhythms

## 1.1 How we adjust to our environment

We humans, like other animals and plants, are constantly fighting against the opposing forces of our environment. Indeed, to fail to do so is to risk death. Long ago, for example, we evolved mechanisms to deal with the problems associated with living on dry land. In this environment, the effects of gravity are no longer balanced by the buoyancy that water provides, and the air, particularly on a dry and windy day, can cause loss of water by evaporation. Yet we can deal with these problems. We also have to cope with a far wider range of ambient (environmental) temperature than do organisms that live in the sea. On land, the sun can provide us with heat in the daytime but, after the sun has set, the air cools much more quickly than the water. In addition we must work continually to ensure that the body receives enough oxygen and food, and removes waste. Aquatic plants and animals, however, do not live in a watery paradise, as they have to suffer the destructive power of wave action and must deal with any changes in the composition of the water. Also, if they live in estuaries or in other bodies of shallow water, there is always the risk of desiccation.

Nevertheless, plants, animals, and humans do survive because they possess sophisticated methods for controlling their internal environment, in spite of the external environmental forces. These methods involve the behaviour, physiology, and biochemistry of the organism.

Consider, for example, the way in which human body temperature is maintained at about 37 °C throughout the 24 hours of the day and the four seasons of the year. In the colder weather we insulate ourselves by wearing more clothes and try to generate more heat by being more active (walking faster and swinging our arms or stamping our feet). Sometimes we shiver and our teeth chatter. In addition, our skin becomes pale because the vessels supplying it with blood have partially closed down. One hormone that is secreted in increased amounts in the cold is adrenaline, which has many actions, one of which is to increase our general metabolism, so producing more body heat. In hot weather, by contrast, we wear less clothing and try to be in less of a hurry — consider the siestas taken around midday in warmer climates. We also release more heat through the skin, as witnessed by its reddening, and lose heat by sweating.

It is important to note that humans are able to control their body temperature by changing their physiology and biochemistry. It is for this reason that they, like birds and other mammals, are said to be 'endothermic' (warm-blooded). For animals that are 'ectothermic' (cold-blooded), their temperature control is mainly behavioural and consists of avoiding temperatures that are too high or too low by hiding in burrows or under stones, for example. For some animals, the demands that are required of them by the environment appear to be too great and they hide from the problem, by spending the winter as a cocoon, by going to sleep throughout the winter (hibernation) or summer (aestivation), or by escaping completely (migration)!

The physiological process of maintaining a variable such as body temperature within far narrower limits than the environment is called 'homeostasis' (from the Greek for 'constant state'), and other homeostatic mechanisms exist to control blood pressure and gases, the intake, digestion, and metabolism of food, and the general composition of body fluids. Homeostasis, and the need for control mechanisms to implement it in living organisms, have been understood for only a little under a century, but we now know that by such mechanisms, changes in the environment — produced by so many different properties of the environment — can be resisted. These mechanisms give organisms a degree of independence of their environment which has, in turn, enabled them to move into other habitats. One example is how, over time, humans have moved out of Africa to inhabit all corners of the globe and to adapt to a vast range of environments.

We are stable creatures in an environment that can be unstable.

## 1.2 A rhythmic world

Although we have outlined some of the problems caused by environmental influences such as temperature, water loss, and gravity, we have not yet considered an all-pervading influence — that of time, which has a profound effect on everything that we do.

We live in a rhythmic world. From early childhood, unless we live in the tropics, we are aware of the annual cycling of the seasons and its effect on nature, from the comparative dormancy in the winter to the activity of summer. There are clear differences in the ambient temperature and amount of daylight between the seasons, accompanied by such as the migrations of birds, the opening of blossom and leaves, the mating seasons, and the waking up from hibernation. It seems clear that such changes, and their importance to farming and to cultural activities, have long been recognized, as shown by the solar alignment of many ancient megalithic monuments. Perhaps the most famous of these is Stonehenge in the south of England, the orientation of which indicates its builders' awareness of the changing position of the sun with the seasons. Other structures whose orientation can be linked to natural seasonal phenomena are found throughout the world.

In addition to these annual changes, there are daily ones. It is these daily changes and the responses that living organisms have evolved to deal with them that are the subject of this book.

The distinction between night and day is one of the most pervasive rhythms that we experience. The Earth spins on its axis, alternately exposing us to, and then hiding us from, the light of the Sun. The time taken for the Earth to complete one revolution has decreased over the aeons of time, but at the moment it is, give or take the odd second, 24 hours. This time span defines the solar day. Many aspects of our environment show a rhythm with a period of 24 hours as a result of this solar day, light and temperature being the most obvious. (The period of a rhythm is the time to complete one cycle.)

The Earth also has one natural satellite, the Moon, which exerts a powerful influence upon the water in the oceans. The Moon's gravity attracts this water, as a result of which, two huge 'bulges' of water appear in the oceans, one facing towards the Moon and the other facing away from it. The spinning of the Earth causes these bulges to travel round it, and this gives rise to our high tides. The tides get later by about 40 minutes each day because of the movement the Moon makes around the Earth. Only occupants of the ocean floor or the deepest recesses of caves will escape such rhythmic influences and live, instead, in environments where the passage of daily time goes unnoticed or the tides have no effect. Even here, however, some annual changes might be seen, such as the amount of detritus that showers down from the ocean surface or the types of animal sheltering in the cave.

## 1.3 Daily rhythms in living organisms

It is not only the environment but also the animals and plants which inhabit it that display rhythms. This can be seen most clearly when considering their behaviour. Many animals are diurnal — that is, they are active in the daytime and inactive at night — whereas nocturnal creatures show the opposite orientation. In addition, particularly in hotter regions, there are animals that are only active crepuscularly, that is, about the times of dawn and dusk, when it is neither too hot nor too cold.

We are all aware to some degree of these daily rhythms that surround us. There are few people, for example, who have not woken on some occasion to the sounds of the dawn chorus. In the afternoon, birds become quieter, while butterflies and other insects are more active, searching for sources of nectar. At twilight, mosquitoes and midges appear and then, as darkness falls, nocturnal creatures such as hedgehogs, owls, bats, and many rodents take over. Flowering plants also respond to their environment as is shown by the charming example of Linneaus' flower clock (Table 1.1). These responses of flowers affect the habits of nectar-seeking insects and alter the types of pollen that are being spread by the wind at a particular time of day. Also, if the day is sunny, as drying proceeds, clouds of wind-borne seeds will be released or pods will open, often scattering their contents explosively.

**Table 1.1** Linnaeus' Flower Clock, relating events to the time of day

| | |
|---|---|
| 06:00 | Spotted Cat's Ear opens |
| 07:00 | African Marigold opens |
| 08:00 | Mouse Ear Hawkweed opens |
| 09:00 | Prickly Sowthistle closes |
| 10:00 | Common Nipple Wort closes |
| 11:00 | Star of Bethlehem opens |
| 12:00 | Passion Flower opens |
| 13:00 | Childing Pink closes |
| 14:00 | Scarlet Pimpernel closes |
| 15:00 | Hawkbit closes |
| 16:00 | Small Bindweed closes |
| 17:00 | White Water Lily closes |
| 18:00 | Evening Primrose opens |

Carolus Linnaeus (1707–1780)

Humans are normally diurnal, but we are also more flexible than plants and other animals, so that in the modern world many people work at night and sleep during the day. Artificial lighting has enabled us to behave independently of the natural environment. Even so, we normally synchronize our times of sleep, leisure, and meals, if only so that we can meet other people. Night workers, for example, generally adopt a conventionally-timed lifestyle during their rest days.

Accepting that we respond to our environment, whether it is natural or artificial, we can see that these responses produce further changes. We are about two centimetres taller on getting up in the morning compared with when we go to bed the previous night. In the daytime, with the weight of our body supported by the spine, normal activity, in an upright posture, causes a loss of stature (referred to as 'shrinkage'). During sleep, however, our spine no longer needs to support the weight of our body. Consequently, fluid passes into the discs between our vertebrae, causing them to swell slightly. It is this swelling that makes us 'grow' taller overnight. When we get up, the weight of our bodies compresses these discs and squeezes out the extra fluid. Normally, half of this 'extra' height is lost in the first hour after getting up but, if exercise is taken on waking, then the loss of height is more rapid (Fig. 1.1).

These results suggest that the changes in height are due to the sleep–wake rhythm — and the associated alternation of rest and activity — which, in turn, is a response to our rhythmic environment.

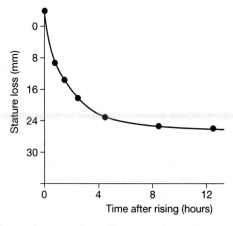

**Figure 1.1** Loss of stature after waking up and resuming an upright posture.

Support for this idea comes from the following observations:

- If sleep is taken during the daytime, then height is gained at the same rate and for the same reason as during the night.

- If a person stays awake at night and does not lie down, then the increase in height does not occur.

Plants and animals in general also show physiological and biochemical changes that appear to result from their responses to their environment, but is this the whole story? Is a living creature rhythmic wholly because it is driven by a rhythmic environment and lifestyle? If this were the case, then the response to time in the environment would be analogous to the responses to gravity, to water loss, and to temperature, for example.

But it is not the whole story.

## 1.4 Rhythms from within

Rhythms do not result only from our habits and environment however. To understand why this is so, more detailed measurements of physiological or biochemical processes are required. Figure 1.2 shows two examples — of body temperature and urine flow.

Clearly, temperature and urine production are both higher in the daytime than at night. The issue, however, is whether these results are merely reflections of our environment and lifestyle. For example, it could be argued that higher body temperature in the daytime is due to greater physical activity when it is light, since the extra muscle effort produces an increased amount of heat. In addition, environmental daytime temperatures would be higher, and so heat loss from our body would be more difficult, whilst heat gain from the sun would be greater. All these factors would be reversed at night, when we are asleep and inactive in a cooler and less interesting environment, and body temperature would fall as a result. A similar argument could be applied to urine production, except that the emphasis would be on fluid intake. At night, urine flow would be low because we cannot drink when we are asleep, and in the daytime our kidneys would respond to the increased fluid intake.

Our personal experience, however, indicates that, unlike the case of height, these rhythms do not result wholly from our lifestyle and the environment. Consider the following examples:

- With regard to thermal comfort (whether we think our environment is too hot or too cold), a particular room temperature might

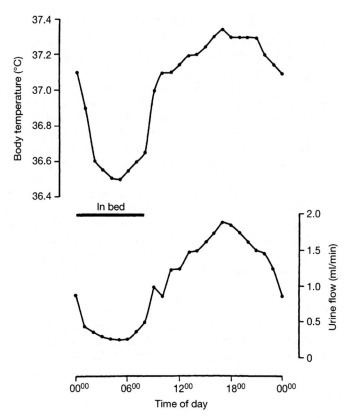

**Figure 1.2** Hourly records of body temperature and urine flow in humans living a normal lifestyle.

make us feel cold in the morning, and yet the same room temperature will sometimes feel hot and stuffy in the evening.

- A bedtime drink does not waken us with a full bladder during the night, whereas a similar drink in the daytime causes urine flow to increase quite rapidly.

- After a flight to a new time zone our lifestyle can become disrupted — we cannot sleep at night, we feel tired during the daytime (at a time corresponding to night in the time zone we have just left), and we lose our appetite for food.

- If we work at night, at first we have difficulty sleeping during the day.

- If, for some reason, we stay up all night (see Fig. 1.3, for example), we feel more tired as the evening and the night wear on. This is intuitively what we would expect and we would (quite rightly) attribute it to the lack of sleep. However, from about 05:00 in the morning onwards, changes occur that do not accord with this view. During the latter part of the night, in spite of still having had no sleep, feelings of fatigue begin to diminish. Indeed, by about the middle of the morning we feel surprisingly alert — we might even not bother with sleep after all since we appear largely to have overcome the effects of the lost night's sleep. By contrast, later the next evening, we will find that our fatigue increases very markedly and we will feel much more tired than on the previous evening. As is indicated in Fig. 1.3, how tired we feel seems to be determined not only by how long we have been awake but also by a rhythmic change superimposed upon this, one that decreases our fatigue at some times (in the daytime) and increases it at others (in the night).

Observations such as these suggest that there are rhythmic changes in our bodies which are produced internally, in addition to effects directly due to our environment and lifestyle. These internal rhythms are generated by a 'body clock' (though this tells us nothing about where it is or what its properties are). Most of the rhythms we shall

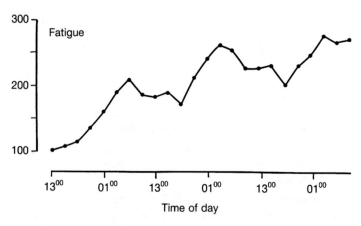

**Figure 1.3** Three-hourly ratings of fatigue (arbitrary units) in a group of soldiers staying awake for three consecutive days.

consider in this book are caused by a combination of both this body clock (an internal cause) as well as external causes (such as changes in the environment). We shall consider the importance to us of these rhythms and how they are produced by a combination of the effects of the body clock, our lifestyle, and the environment.

## 1.5 Separating the internal and external causes of body rhythms

Accepting that the rhythms of body temperature and urine flow are caused by a mixture of internal and external factors, we need some experimental means by which these two factors can be separated. This has been achieved by a 'constant routine' protocol. This protocol entails studying individuals under circumstances in which the external causes of rhythms have been removed. First, the environment must not be rhythmic, and so individuals are studied in constant lighting, humidity, and temperature. Second, the lifestyle must also show no rhythmicity. To achieve this end, individuals remain awake and seated for at least a full 24 hours (the time for the rhythm to show a full cycle) and the way they spend their time is as constant as possible. In practice, time is passed by reading, writing, listening to music, playing cards, doing jigsaws, and so on. Meals also have to be modified so that they do not produce a rhythmic input to the body. One way is to arrange that an identical snack is taken every 3 hours throughout the day and night.

What happens to the daily rhythms of body temperature and urine flow during such a protocol? Figure 1.4 shows that daily rhythms exhibit a similar pattern (with regard to general shape and timing) to when the individual was living and sleeping normally, although their amplitude (the difference between maximum and minimum values) is decreased. Thus the rhythm that persists during a constant routine protocol must arise internally. It is produced by the body clock.

Figure 1.4 also tells us more about the rhythms. First, the differences between the rhythms under normal conditions (when the environment is rhythmic and we are living a rhythmic lifestyle of sleep and activity) and under the constant routine conditions must be due to the environment and our lifestyle — in other words, to external causes. Comparing the amplitude of the rhythms under the two conditions gives us some idea of the relative importance of internal and external causes. It can be seen that the two causes appear to be of similar importance in the

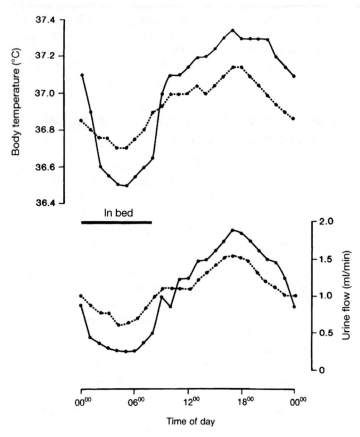

**Figure 1.4** Hourly records of body temperature and urine flow in humans living a normal lifestyle (solid line) and on a 'constant routine' (dashed line). (Time spent in bed applies to a normal lifestyle only.)

case of body temperature (Fig. 1.4, top) but the internal cause in the case of urinary flow (Fig. 1.4, bottom) is rather smaller. Secondly, a comparison between the rhythms obtained normally and in constant routine conditions indicates that they are similarly timed. That is, during the daytime, the body clock as well as our lifestyle and environment all raise body temperature and increase the production of urine. By contrast, at night, the body clock, sleep, and our inactive environment act together to decrease both body temperature and urine production.

It is important to note that if the timing of the external causes is suddenly changed, as after a transatlantic flight or after starting night work, the internal and external causes will no longer be aligned. This mismatching of the internal and external causes of a rhythm can produce difficulties, which will be considered in detail in Chapters 10 and 12.

## 1.6 'Larks' and 'owls' — the interaction between internal and external causes

Although individuals are similar in that they have many rhythms in common, composed of external and internal causes, there are also differences between them. Some of the best known are the differences between 'larks' and 'owls'.

'Larks' (or morning types) are those people who tend to wake up and get up early. They welcome the new day and tend to feel at their best, and so do more important things, in the morning. Evenings are their weakest time since they feel fatigued and want to go to sleep relatively early. 'Owls' (or evening types) are the opposite in all of these respects. They stay up late, are slow to get going in the morning, and seem to gain in strength as the day wears on. They prefer to leave important matters until the afternoon, and by the evening they are still full of energy (at a time when the larks are beginning to wilt).

It would be misleading, however, to think of everybody as either a lark or an owl. There is a wide range between the two extremes of behaviour preference, and most people can be defined as 'intermediate'. A brief test at the end of this chapter has been designed to give some idea of how much of a lark or owl you are.

Why are some of us larks and others, owls? Is it a function of our body clock or our lifestyle, or due to some interaction between the two? Just as we differ in mental and physical characteristics, so, too, we differ when the detailed properties of our body clock are considered. There is evidence that the body clocks of larks run slightly faster than the average of the population as a whole, so that there is a tendency for the effects of the clock (including getting to sleep and waking up) to occur slightly earlier in them. Since owls have body clocks that run a little more slowly than average, there are tendencies for later bedtimes and 'lie-ins'.

In addition, circumstances can cause us to behave like larks or owls. Thus, people whose work necessitates an early start (for example, farmers, bakers, some factory workers, and long-distance commuters) will routinely time their lifestyle (retiring, rising, and meal times) earlier than average and their environment — when it is dark or light, noisy or quiet — will be adjusted to match. By contrast, some sections of the community (for example, students, disc jockeys, and restaurant and pub workers) plan their lifestyles and environment to be later than average. Such a lifestyle and, with it, the timing of daily rhythms, often continues into retirement — due to habit rather than need.

Whether one is a lark or an owl can depend upon the properties of our body clock as well as the timing of our lifestyle. Sometimes the two reinforce each other. An owl will often pursue hobbies or interests late into the night, and the preferred lifestyle and body clock will accentuate one another. In the same way, larks might avoid late-night parties in favour of early morning exercise. It might therefore be imagined that an individual with a lark-like clock but an owl-like lifestyle, or vice versa, would have difficulties. This is indeed the case and will be covered when shift work is considered in Chapters 12 and 13.

## Now try this — are you a 'lark' or an 'owl'?

For each of the following questions choose the most appropriate answer. *Do not cross-check your answers.*

**1.** *What time would you choose to get up if you were free to plan your day?*

A    05:00–06:00
B    06:00–07:30
C    07:30–10:00
D    10:00–11:00
E    11:00–12:00

**2.** *You have some important business to attend to, for which you want to feel at the peak of your mental powers. When would you prefer this meeting to take place?*

A    08:00–10:00
B    11:00–13:00
C    15:00–17:00
D    19:00–21:00

**3.** *What time would you choose to go to bed if you were entirely free to plan your evening?*

A    20:00–21:00
B    21:00–22:15

C    22:15–00:30
D    00:30–01:45
E    01:45–03:00

**4.** *A friend wishes to go jogging with you and suggests starting at 07:00–08:00. How would you feel at this time?*

A    On good form
B    On reasonable form
C    You would find it difficult
D    You would find it very difficult

**5.** *You now have some physical work to do. What time would be best for you?*

A    08:00–10:00
B    11:00–13:00
C    15:00–17:00
D    19:00–21:00

**6.** *You have to go to bed at 23:00. How would you feel?*

A    Not at all tired, unable to get to sleep quickly
B    A little tired, but unlikely to get to sleep quickly
C    Fairly tired, and likely to get to sleep quickly
D    Very tired, very likely to get to sleep quickly

**7.** *When you have been up for half an hour on a normal working day, how do you feel?*

A    Very tired
B    Fairly tired
C    Fairly refreshed
D    Very refreshed

**8.** *At what time of the day do you feel best?*

A    08:00–10:00
B    11:00–13:00
C    15:00–17:00
D    19:00–21:00

**9.** *Another friend suggests jogging at 22:00–23:00. How would you now feel?*

A    On good form
B    On reasonable form
C    You would find it difficult
D    You would find it very difficult

Now, score your answers. Add up the points for the 9 questions:

| Question 1 | Question 2 | Question 3 |
|---|---|---|
| A = 1 | A = 1 | A = 1 |
| B = 2 | B = 2 | B = 2 |
| C = 3 | C = 3 | C = 3 |
| D = 4 | D = 4 | D = 4 |
| E = 5 |  | E = 5 |

| Question 4 | Question 5 | Question 6 |
|---|---|---|
| A = 1 | A = 1 | A = 4 |
| B = 2 | B = 2 | B = 3 |
| C = 3 | C = 3 | C = 2 |
| D = 4 | D = 4 | D = 1 |

| Question 7 | Question 8 | Question 9 |
|---|---|---|
| A = 4 | A = 1 | A = 4 |
| B = 3 | B = 2 | B = 3 |
| C = 2 | C = 3 | C = 2 |
| D = 1 | D = 4 | D = 1 |

## Interpreting your score

Your score can range from 9 to 38. This is only a guide, of course, but your score can be interpreted as follows:

| 9–15 | Definitely a lark |
| 16–20 | Moderately a lark |
| 21–26 | Neither a lark nor an owl — an intermediate type |
| 27–31 | Moderately an owl |
| 32–38 | Definitely an owl |

Most would score between 21–26; only about 5% of the population would score 15 or less and another 5%, 32 or more.

# Chapter 2

## Some properties of the body clock

## 2.1 Limitations of constant routines

Even though the role of the body clock as the internal cause of a rhythm can be investigated by performing a constant routine, this routine lasts only for 24 hours or so, and so enables us to assess the effects of the clock over the course of one cycle only. We need to study the clock for much longer periods of time to assess its ability to keep time, for example. Experiments in which volunteers have been deprived of sleep are fairly common and have been used to establish how an individual's physiology, physical performance, and, particularly, mood and mental performance, respond to sleep loss. From the present viewpoint, such experiments are of very limited use since subjects have continued to be exposed to the alternation of day and night and to other rhythmic inputs.

What would be more valuable is an experiment in which volunteers remain on a constant routine for as many days as possible. Not surprisingly, such experiments are rare! A Swedish study went some way towards achieving this aim by using soldier 'volunteers' and keeping them awake for about 72 hours in a constant environment. Subjects performed a series of tests and took an identical snack every 3 hours throughout the study. The results showed that rhythms with a period of 24 hours continued, and this confirmed the activity of the body clock during the experiment (Fig. 2.1). As the experiment wore on, however, there was clear evidence that loss of sleep was beginning to dominate the results. There was a progressive increase in fatigue (tiredness) and decrease in speed of shooting at a target.

Therefore, we cannot use this method to obtain clear information about the body clock, because other factors begin to intrude. In any case, investigating only three cycles of a clock is still inadequate if we are to determine in detail its accuracy and the factors that affect it.

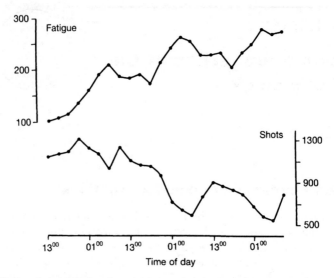

**Figure 2.1** 3-hourly ratings of fatigue (arbitrary units) and speed of firing at a target (high score indicates high speed) in a group of soldiers staying awake for 3 consecutive days.

Since an even longer constant routine is not feasible, a different type of experiment is required.

## 2.2 Studying the body clock for extended periods of time

Consider the following experimental design for studying a sole volunteer. He or she is instructed to go to bed when feeling tired, to get up when feeling rested, and to eat meals when it is estimated (from feelings of hunger or from what has been done so far during the waking period) that it is breakfast, lunch, or dinner time. Snacks and any (indoor) exercise can also be taken when and if desired. In such an experiment, how any time is spent is entirely at the discretion of the volunteer. The 'catch' is that the volunteer has no cues from clocks, television, or other humans as to what the time really is — these are all excluded from the environment. The experimenters keep a record of what the volunteer does and, by means of a hidden clock, when it is done. Because there are no constraints placed upon the volunteer's activities in such a time-free environment, these are called free-running experiments. The bodily rhythms measured during them are known as free-running rhythms.

What happens? Do individuals lose all sense of time and become erratic in their habits. Alternatively, do they maintain a rhythmicity in their lifestyle and bodily functions?

In fact, individuals do not become random in their habits. Rhythms of body temperature and the elimination of waste substances in the urine, and whether the volunteer is asleep or awake and active (the sleep–wake cycle), all continue. This is shown in Fig. 2.2 in the case of a free-running experiment lasting nearly 4 weeks.

Notice that the times of waking become a little later each day and this indicates that the length of the sleep–wake cycle is greater than 24 hours — about 25 hours in the case shown. It is important to realize that the general pattern of sleep and activity, as of the other rhythms, is very regular. It is also important to remember that, as far as the volunteer is concerned, rising from bed takes place at the 'normal' time each 'morning', in order to plan a full day ahead. For the moment, the key point to be gained from these results is that the maintenance of rhythms, in the absence of external information about time, confirms the presence of an internal body clock. (The implications of

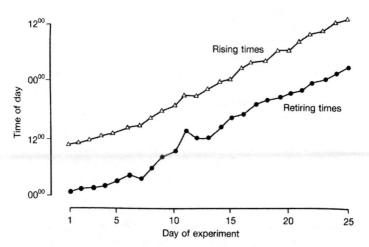

**Figure 2.2** Times of retiring and rising on successive occasions in an individual isolated from all knowledge of the passage of time — a free-running experiment.

a body clock running slow with a period (cycle length) greater than 24 hours, rather than exactly 24 hours (the same as the solar day) will be considered later.)

Before continuing with our main theme, there are a few points that need further discussion; they relate to what happens in time-free conditions, and whether alternatives to the body clock can be offered as explanations.

## 2.3 Time-free environments

The first problem is finding a time-free environment in which free-running experiments can be performed. One possibility is to perform experiments in the relatively constant conditions that exist near the Poles. In summer, the sun does not set and so there is no alternation of day or night to give cues as to the passage of time. There are also likely to be few time cues from animals or other humans, such as tourists. The continuous darkness during the winter could also be made use of, but this has been less attractive, both to volunteers and to experimenters — such an environment might be free of time cues but is inhospitable because of its climate. There are also potential problems if emergencies arise or if there is failure of the monitoring apparatus — and in sub-zero temperatures, this is a real possibility. One advantage of such a site, however, is that samples of blood and urine that are required for analysis of rhythms are readily refrigerated so that their deterioration is less of a problem than in other environments.

A second time-free environment is provided by underground caves. In these there is no penetration of natural daylight and the only sounds are those of running and dripping water. Volunteers have stayed alone in such environments for more than nine months. Clearly they are easier 'laboratories' to reach and far less dangerous than the Poles with regard to climate and the hazards of survival. Even so, they are uncomfortable and unpleasantly damp and dark, and the need for artificial lighting and heating can produce fumes that add to the inconvenience of the situation.

A third, and the preferred, method is to build an isolation unit, specially constructed for such experiments. (We stress that 'isolation' in this context implies isolation from time cues in the environment rather than sensory deprivation.) The first one was built in Germany and later, in 1968, one was built in Manchester, England. Subsequently, several others have been constructed in different parts of the world, differing in their sophistication and the type of experimental measurements that can be taken of the volunteers living in them. A diagram of the new unit at Liverpool John Moores University is shown in Fig. 2.3.

The advantage of such experimental facilities over the Poles or underground caves is that they can combine some of the comforts of home with the more exacting requirements of free-running experiments. Also, there is the possibility of taking many more measurements in such a laboratory than in experiments at the Poles or in caves.

In spite of such differences, however, it is reassuring to know that experiments performed in these three rather different time-free environments have all given results like those shown in Fig. 2.2.

## 2.4 Alternatives to the concept of a body clock

Another major problem that must be considered is the possibility that the continuing rhythmicity found in free-running experiments is not due to a body clock. Two alternative explanations, and comments on them, follow.

One possibility is that the subject is responding to an external rhythmic influence that has not been controlled. The problem is to decide what such an influence might be. Influences upon humans of the planets, the moon, and factors such as magnetic fields, atmospheric pressure, and cosmic rays have been imagined by some, including lovers, astrologers and those who, in the past, have diagnosed types of 'lunacy'. However, when such influences are considered as explanations of the results of free-running experiments, the following problems have never been satisfactorily dealt with:

- Where are the sense organs that pick up such external factors? There is evidence for a magnetic sense in some animals, including humans, but its role in influencing these rhythms — let alone actually being responsible for them — is not at all established. Sense organs for lunar and planetary influences, for atmospheric pressure, and cosmic rays are, as yet, purely hypothetical from a scientific viewpoint.

- Why do individuals have free-running rhythms that differ in period length? That is, free-running experiments have shown that the

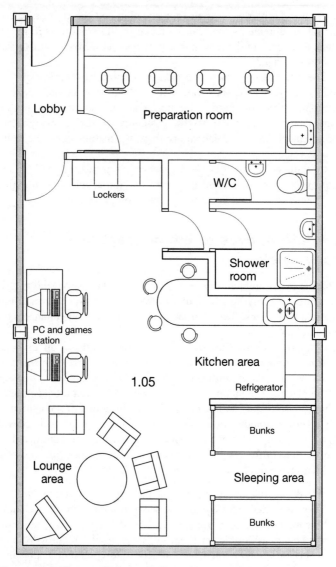

**Figure 2.3** The new isolation facility at Liverpool John Moores University.

period of the rhythms in different individuals varies somewhat but is generally within the range 24–26 hours. (As described in Chapter 1, 'larks' tend to be at the lower end of the range and 'owls' towards the top of it.)

An alternative explanation is that the results are due to a regular structuring of an individual's lifestyle and habits rather than a reflection of the output of some body clock. There is no doubt that individuals tend to structure their day around routines of meals, personal hygiene, leisure, work, and so on. However, to suggest that this will account for the results observed in Fig. 2.2 raises some problems.

Firstly, it is not clear how the duration of sleep could be controlled as regularly as is observed to be the case. Even though it might be argued that a regular lifestyle implies a regular degree of fatigue, and so will require a regular amount of sleep for recuperation, in practice, some irregularities are also present. Individuals go to bed later or earlier than average on certain occasions, as can be seen by close inspection of Fig. 2.2. For example, an individual might go to bed later than usual on one occasion because a piece of work or some leisure activity took longer than normal. It would be predicted that this will make the individual more tired than usual, and so we might guess that sleep would be longer than average. In other words, an individual's sleep–wake cycle would run more slowly on interesting days. Conversely, it would be predicted that the cycle would run more quickly on days when the individual was bored (because he or she would go to bed earlier due to the lack of something interesting to do, and so require less sleep).

This idea — that longer activity spans would be followed by longer sleeps and vice versa — does not find experimental support. If anything, the reverse is seen: long 'daytimes' tend to be followed by shorter sleeps, and shorter 'daytimes' by longer sleep (consider days 7 and 11 in Fig. 2.2, for example). This result is easier to explain if we assume that a body clock is responsible for the alternation between sleep and activity, and that such a clock will wake an individual when a certain stage of the sleep–wake cycle is reached. In such a case, after a late bedtime, this stage will be reached after a **shorter** period of time than usual.

In practice, however, a later bedtime is followed by a **slightly** later time of waking, even though the sleep has not been long enough to compensate completely for the late night. This indicates that fatigue does, after all, exert some influence upon the length of time we sleep.

Clearly the factors that determine the amount of sleep we take is a complex issue, and one we will return to in Chapter 6. In the present context, the important point is that a longer time awake does not increase sleep in the same proportion.

The second problem with suggesting that regular structuring is responsible for the results in Fig. 2.2 is that when all results from the free-running experiments are considered, the sleep–wake cycle is greater than 24 hours. If it were due to some 'memory' of our lifestyle then we would predict that about half the population would show a value less than 24 hours and the free-running periods of individuals would be distributed fairly symmetrically about an average value of 24 hours. However, it is found in practice that even 'larks', whose body clocks might run faster than the average for the population as a whole, rarely show a period less than 24 hours in free-running experiments.

## 2.5 Long-term free-running experiments

The regularity that is observed in free-running experiments, and which has been interpreted as evidence for the body clock, has been stressed so far. However, it should be added that, particularly when experiments lasting several months are performed, certain irregularities of the sleep–wake cycle begin to appear. Occasionally, a sleep is missed altogether or it is much longer than usual. Sometimes, volunteers even adopt a sleep–wake cycle that lasts about twice as long as normal — maybe about 50 hours. In such circumstances, rhythms of food intake tend to follow the altered length of the waking period, as a result of which the length of time between waking and eating breakfast, between breakfast and lunch, and so on, all tend to increase in proportion to the lengthening of the waking span. This result suggests that our habits and lifestyle do, after all, play some part in determining when we eat, but the interesting thing is that, even in these circumstances, the rhythm of body temperature (and of many other variables) retains its regular period of just slightly greater than 24 hours.

These findings have several important implications. First, they do not require us to dispense with the idea of an internal body clock, if only because they appear very rarely in experiments lasting only a week or so — in contrast to the rhythms with a period just greater than 24 hours, which are regularly seen. Instead, the results suggest that the body's timing system needs an occasional rhythmic input from the outside world to run smoothly. The alternatives to an internal body clock (planetary influences, cosmic rays, and so on) do not offer an explanation of these irregularities. Second, the results indicate that some rhythms (body temperature, for example) are connected more closely to the body clock than are others (the sleep–wake cycle, for instance) and that, when the body clock is not responsible for a rhythm, then environmental factors and habits are more important. The third implication of the sleep–wake cycle running with a period that is different from that of the body temperature rhythm is that there will be times when sleep will coincide with the peak of the body temperature, and that there will be others when activity will coincide with the trough. These results are very different from those found normally, when sleep coincides with the trough of the body temperature rhythm and waking, with its peak (see Fig. 1.2). Implications of this, particularly with regard to jet lag and shift work, are considered later (see Chapters 10 and 12).

Some details of the site and nature of the body clock are to be found in Chapter 8. At this stage, it is sufficient to know that we possess one, wherever it might be.

## 2.6 Adjusting the body clock

Even accepting the overwhelming evidence that we possess a body clock, we seem to possess one that is of little use, since it tends to run with a period different from that of a solar day! Indeed, it is this observation that leads to it often being referred to as a 'circadian' clock (from the Latin 'circa' — about; 'diem' — a day) — that is, one with a period that only approximates to 24 hours. In the cave experiments described above, the average period of the clock was around 25 hours. Such a body clock, if correctly timed one day, would be approximately 1 hour late the next, about 2 hours late the day after that, and so on until it would be 'useful' again only after about 3–4 weeks. This can be readily appreciated when the rising times of the volunteer in Fig. 2.2. are considered. By day 9 of the experiment, the volunteer showed a timing of the sleep–wake cycle that was more appropriate for a night worker, rising at about 09:00 and retiring at about 18:00 by solar time (the time we normally use). It should be stressed though that by 'body time', the volunteer continued to rise at about 'midnight' and rise about '08:00', just before eating breakfast. Also, by day 22, the volunteer was, once again, retiring and rising at what an external observer would call 'normal' times, though the volunteer had by now lost a whole day, having retired and risen only 21 times.

Clearly this pattern is not useful and, equally clearly, this is not what really happens since, if this were the case, then daily rhythms would be observed to be timed irregularly when different individuals were compared. The results shown in Fig. 1.2 are routinely found in individuals living normally. It is stressed that the body rhythms are similarly timed on successive days in any individual, and in the different individuals measured on the same day. Even in 'larks' and 'owls', in whom there are differences in the timing of rhythms compared with the majority of the population (the 'intermediate' type), these differences are of a few hours only and are reliably found from one day to the next. This means that the circadian rhythms, and the body clock that produces them, must be adjusted so that they are in phase with the solar day.

## 2.7 Adjusting the body clock — zeitgebers

How is this adjustment achieved? When we adjust our watch, it is by synchronizing it with an external signal, often taken from the radio or television. It would be possible, though more tedious, to take the signal from some natural phenomenon (for example, when the sun or another star reaches its highest point in the sky). The same principle — that of using some external reference time — is used by plants and animals, including humans, for adjusting their body clock. The external reference time that is used biologically is the timing of one or more rhythms in the environment.

The natural rhythms that are important depend upon the environment, the organism, the stage of life history that has been reached, and the period to which the body clock will be adjusted. The environmental rhythms that are giving time cues to the organism are called zeitgebers (from the German 'Zeit' — time; 'geber' — giver). Some zeitgebers are given in Table 2.1.

When the period of the body clock has been adjusted to equal that of the zeitgeber, it is said to have been 'entrained' by that zeitgeber. The body clock of most animals and plants is entrained to a period of 24 hours, equal to that of the solar day. However, attention is drawn to those creatures that inhabit the inter-tidal zone, namely the region of the shore between high and low tides. Their lifestyle is dominated by the tides rather than the sun, and they are synchronized with

**Table 2.1** Some organisms and the time cues that adjust their body clocks

| Organism | Time cue is the rhythm in: | Adjusts the period of the clock to: |
| --- | --- | --- |
| Many plants and herbivores | Light and darkness | 24 hours |
| Predatory animals | Food availability | 24 hours |
| Creatures in the inter-tidal zone | Buffeting by waves and immersion in water | 12.4, 24.8 hours |
| Newborn rodents | Mother's activity | 24 hours |
| Some bats | Starlight | 24 hours |
| Lizard (cold-blooded animals) | Environmental temperature | 24 hours |
| Rat fetus | Mother's hormones | 24 hours |

(entrained by) the tides rather than the sun, their cycle length being 24.8 hours.

It has already been described how living organisms respond to their environment, and how this environment is the external cause of a measured rhythm. It can now be seen that this integration between the environment and the organism is even more intimate, with rhythms in the environment also adjusting the timing of the body clock.

## 2.8 Zeitgebers in humans

Humans are likely to be 'special', since they are less dependent on the natural environment than are plants and other animals. (Individuals living 'closer to nature' in pre-industrial societies have been studied hardly at all.) Studies of humans indicate that they are entrained to the 24-hour solar cycle, however, and so time cues must exist, albeit often artificial ones. Since the circadian rhythms seem to be in phase with the sleep–wake cycle (see Fig. 1.2), it seems reasonable to consider if it is this alternation between sleep and wakefulness that is acting as the zeitgeber. However, the sleep–wake cycle automatically exposes us to a wealth of possible zeitgebers, any or all of which might be important.

Consider the following scenario. There are times when it is convenient for us to go to sleep — the shops might not be open, we have no appointments, or our colleagues and friends might not be around.

There will also be times when we are expected to be quiet rather than to go outside and mow the lawn. When we choose to sleep, we are in a quiet and dark environment, and we do not eat. The times we are awake are often imposed upon us by social factors — the needs to work, to shop, to fulfil appointments, and so on. In practice, of course, these competing demands are generally met by sleeping at night and being active in the daytime. In other words, we act as diurnal creatures and, once again, this shows the biological abnormality of night work and daytime sleep.

As a result of this timing of our sleep–wake cycle, it can be seen (Table 2.1) that we become exposed to many potential zeitgebers. In practice, attempts have been made to separate the important ones as far as humans are concerned. The evidence indicates that the roles of eating and fasting, and of mental, social, and physical activity and inactivity are much weaker zeitgebers in humans than is the rhythm of the light–dark cycle, even when the light is artificial and comparatively dim.

Under normal circumstances, the potential zeitgebers act in harmony to entrain our daily rhythms to a solar (24-hour) day. This results in the peak of body temperature being near the middle of daytime activity and its trough near the middle of nocturnal sleep. This description acknowledges that we have achieved an independence from the natural environment because, unlike other animals, we do not have to adjust our activities to coincide exactly with the sun or the tides.

The detailed timing of exposure to the zeitgebers that adjust our body clock and, therefore, the exact timing of the rhythms produced by this clock, will depend upon social factors and our work. This is illustrated by the comments made earlier about differences between 'larks' and 'owls'. As further examples of these differences, compare the lifestyle of university students — whose tendency to socialize in the late evenings results in fairly late bedtimes and rising times — with that of farmers, who have to deal with livestock and are thus required to be active earlier. Then there are people working in or frequenting nightclubs, whose time cues are particularly late. Extreme examples exist when night workers are considered, but these are so extreme that they give rise to special problems that will be dealt with later (Chapter 11).

As is indicated by the examples of 'larks' and 'owls', individuals can alter their lifestyle and, with it, their exposure to the zeitgebers.

Some people who have fewer external commitments because they are single, unemployed, or self-employed, for example, might even ignore the zeitgebers almost completely, and their rhythms might begin to lose entrainment and become free-running. An example of this can be found in some individuals who live temporarily at or near the Poles. These are abnormal environments as far as time cues are concerned, since there are extended periods of continuous light or darkness, around the summer and winter solstices, respectively. In such circumstances, individuals need to generate an artificial light–dark zeitgeber that complements the sleep–wake cycle and its concomitants (activities, feeding, and so on) if they wish to be entrained rather than free-run. For those who do this — and individuals who work communally have no choice in the matter — rhythms remain entrained to the zeitgeber period, generally one with a period of 24 hours. By contrast, for those who work alone, the need for entrainment can be less or even non-existent, and such individuals might choose not to respond to, or be exposed to, zeitgebers. Then they will show typical free-running rhythms. When extra commitments arise, particularly when others are involved, such a lifestyle has to be modified.

To some extent the majority of us behave in this way at the weekends or during holidays. At such times, we are less constrained by timetables and we might have a fuller social life in the evening. We will tend to go to bed later and have a morning lie-in. All of this causes a slight delay of our body clock. The problem arises on Sunday night (or at the end of the holiday) when we have to go to bed 'early' (relative to the time of our body clock) and so will not be ready for sleep. In addition, we will be woken 'early' by the alarm the next morning, and might experience 'Monday morning blues'.

## 2.9 Adjusting your own body clock

Sometimes we want to change the timing of our lifestyle. For example, after a series of late nights and lie-ins (on holiday perhaps), it might be time to get back to our normal hours of sleep and work. Also, on retirement, we might wish to alter a lifetime of rising early in the morning and retiring early at night. It becomes a matter of changing the timing of good quality sleep. Because, as has already been mentioned and will be discussed in more detail in the next chapter, sleep is much influenced by the timing of the body clock, we need to adjust the 'zeitgeber package', particularly the light–dark cycle that controls the timing of our body clock. In later chapters we will consider how to adjust our body clock to the larger shifts — as after a time-zone transition or during night work; here, we consider smaller modifications.

## Example 1: becoming less of an 'owl' or more of a 'lark'

This is a possible problem after holidays or long weekends; you do not feel tired at the new earlier bedtime but do feel tired when it is time to get up. One method is to make the following adjustments — and to stick to them for a few days while the body clock follows the lead given by your altered lifestyle. The method is designed to accentuate ways that advance your body clock and play down ways that cause it to delay.

- Make evenings a time of relaxation. **Do not** go jogging, but **do** relax in dim lighting and have a warm (not hot) soak in the bath.

- If you cannot get to sleep, **do not** get up or have a meal or snack, but **do** relax and read quietly instead. Remember that sleepiness is promoted by a warm, quiet environment.

- When it is time to get up, **do so** instead of pulling the bedclothes over you. **Do not** take naps in the daytime if you feel tired — to do so makes you less tired the next evening. If you resist any temptation to nap in the daytime you will be more able to get to sleep at your next bedtime.

- If possible, **do** take a brisk walk or gentle exercise first thing in the morning, particularly if it is light then. (This advice has been found to be beneficial for athletes preparing for morning sports events; the issue of harder exercise, such as jogging, will be covered in Chapter 4). **Do not** lounge around feeling sorry for yourself.

- **Do** try to concentrate your activities in the first part of the day and **do not** try to arrange anything strenuous, mentally or physically, late in the day.

- **Do** try to advance your meal times by the same amount as you advance your times of getting up and going to bed.

## Example 2: becoming more of an 'owl' or less of a 'lark'

This desire towards becoming more of an evening type can occur when, for example, you no longer need to get up early for work.

You wish to get up and go to bed later, but tend to wake up too early and to become tired too early in the evening. This is quite common in older people who have retired. (Unfortunately, as we shall discuss in Chapter 7, getting unbroken and refreshing sleep can become more of a problem as we get older.) The advice is now designed to promote a delay of your body clock and reduce those factors that might advance it.

- **Do** try to be physically or mentally active in the evening — taking a walk, visiting friends, or playing a game, for instance. **Do not** relax in dim light in front of the television — unless there is something that you find stimulating.

- If you wake up early, **do not** get out of bed but **do** relax instead and 'snooze'. This will be easier, of course, if your bedroom is dark and quiet.

- **Do** try to arrange for your day to start as leisurely as possible; take time over your breakfast — but indoors rather than outside in the light.

- **Do** try to have a nap after lunch; this will reduce your tiredness in the evening.

- **Do** try to arrange for your most active work to be done later in the day.

- **Do** try to change your meal times to fit in with your delayed times of getting up and going to bed.

## Now try this: how strong are your zeitgebers?

The need for zeitgebers to entrain the body clock to a stable 24-hour period has been described, as has the fact that they originate from the environment and that a regular exposure to them is necessary for a regular lifestyle. The diary below is one way to establish how regular your lifestyle and environment are.

Fill in a log (like the example given below) every hour over the course of a 'typical week'. Record separately the categories of activity, lighting, food, and drink.

**Specimen log**

**Day** _____

| Time | Activity | Lighting | Food | Drink |
|------|----------|----------|------|-------|
| Midnight–01:00 | Sleeping | Dark | — | — |
| 01:00–02:00 | " | " | — | — |
| 02:00–03:00 | " | " | — | — |
| 03:00–04:00 | " | " | — | — |
| 04:00–05:00 | " | " | — | — |
| 05:00–06:00 | " | " | — | — |
| 06:00–07:00 | " | " | — | — |
| 07:00–08:00 | Sitting | Artificial | Breakfast | Tea |
| 08:00–09:00 | Walking | Natural | — | — |
| 09:00–10:00 | Sitting | Artificial | — | — |
| 10:00–11:00 | " | " | Snack | Tea |
| 11:00–12:00 | " | " | — | — |
| 12:00–13:00 | Standing | " | — | — |
| 13:00–14:00 | Sitting | Natural | Lunch | Tea |
| 14:00–15:00 | " | Artificial | — | — |
| 15:00–16:00 | " | " | — | — |
| 16:00–17:00 | " | " | — | Coffee |
| 17:00–18:00 | Walking | Natural | — | — |
| 18:00–19:00 | Sitting | Artificial | Dinner | Coffee |
| 19:00–20:00 | Exercise | Natural | — | — |
| 20:00–21:00 | Sitting | Artificial | — | — |
| 21:00–22:00 | " | " | — | — |
| 22:00–23:00 | " | " | Snack | Horlicks |
| 23:00-midnight | Lying | " | — | — |

## Scoring the log

One easy way is to calculate the percentage of possible occasions when a particular activity or environmental condition applied over the course of a whole week; this way, only something that occurred in the same hour on all days would score 100%.

Examples of this kind of analysis as applied to the types of activity are shown below (Fig. 2.4) for:

- a person on a normal routine working in the daytime;
- a night worker;
- somebody with a very irregular lifestyle;
- a newborn baby;
- a watch-keeper on a merchant ship.

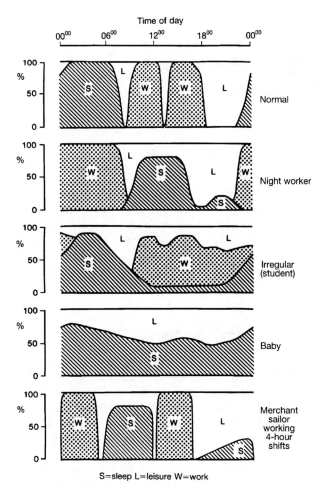

**Figure 2.4** Logs of sleep, work, and leisure, at different times of the day, for individuals living different lifestyles. (For more details of how to produce such a graph, see the text.)

These examples will be referred to and explained later in the book. Advice on strengthening time cues and attempting to make rhythms more regular is given in the 'Now try this' section at the end of Chapter 7.

# Chapter 3

## Fatigue, sleep, and hormones

So far, we have concentrated on the rhythm of body temperature. This is mainly because it is comparatively easy to measure accurately, repeatedly, and cheaply — ideal attributes for research work! Also, it illustrates clearly some of the principles of circadian rhythms. However, those rhythms with more obvious effects are of more general interest. Amongst these, the rhythms of fatigue and sleep are very important.

## 3.1 Fatigue and alertness

The evidence has already been outlined (Chapters 1 and 2) that the rhythms of fatigue (tiredness) and body temperature have both external and internal causes, and that the rhythm of fatigue is the opposite to that of body temperature. (Compare the top part of Fig. 1.4, showing temperatures during a constant routine, with the sensation of fatigue measured in similar circumstances as shown in Fig. 1.3.)

For fatigue, and its opposite, alertness, the evidence has to be based upon subjective measurements whereas, for body temperature, objective measurements are possible. The rhythms of body temperature and alertness are timed very similarly, both showing higher values in the daytime and lower values at night. A detailed explanation of what makes us feel alert is not known, but it is undoubtedly some function of brain activity. Activity within the brain increases with body temperature because enzymes (proteins that are responsible for carrying out the chemical processes within the body) act more quickly as the temperature is raised. This relationship between body temperature and the speed of biological processes applies throughout nature. It is why butterflies warm up before flight, either by exposing their wings to the sun's heat or by vibrating them and generating heat internally; it is why lizards bask in the morning sun; and it is the explanation for the saying: 'Fast runs the ant as the mercury rises'.

If a raised body temperature promotes alertness and staves off feelings of fatigue in the daytime, the fall of temperature in the evening and its low values during the night are equally important for signalling to us that it is time for bed and sleep; high body temperatures at these times would be disadvantageous. We all have experienced occasions when we cannot settle down and get to sleep at night. Many factors can cause this but one of them would be a body temperature that is not falling fast enough in the evening, due to excitement, anger, or apprehension, for instance. It is also a common finding when suffering from jet lag or trying to sleep in the day after night work. In both of these cases (which will be discussed in detail in later chapters), our body clock is wrongly timed for our lifestyle. It does not cool us down and make us feel fatigued and ready for sleep at bedtime.

Figure 1.3 also indicates that fatigue increases in relation to the total time that has elapsed since we last slept. This effect can be seen as an increasing need for recuperation, due to the rigours of being awake — a kind of 'sleep debt'. How tired we feel at any moment is determined by a combination of these two factors. Fatigue feels unpleasant, but the lack of alertness might have disastrous consequences if we are driving a car or undertaking a task where errors might have serious repercussions, such as looking after a nuclear power station.

There has been much interest in trying to model the effects of time of day (the rhythmic component) and time awake (the sleep debt component) on alertness. One of the most successful is illustrated in Fig. 3.1.

In its simplest form, this model postulates that an alertness 'score' can be calculated as the sum of a rhythmic component (factor C, that is parallel to the rhythm of body temperature) and a component that represents the decline in alertness due to time spent awake (factor S). This latter component starts high, after the individual has woken up from sleep, and falls with time awake, at first quickly but then progressively more slowly. This factor recovers during sleep, when it is termed factor S' (Fig. 3.1, top). Use of this model then enables the alertness score under a particular set of conditions (time of day and a time spent awake) to be calculated (the sum of S+C). The bottom of Fig. 3.1 shows the changes in alertness at the end of a normal day and after a first night shift. On waking, subjects are on the '0h' curve (no time awake).

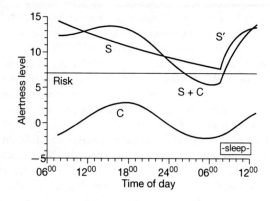

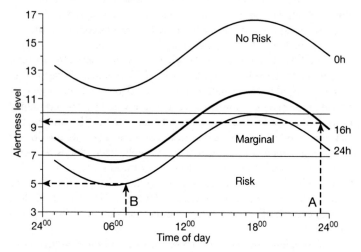

**Figure 3.1** Top: the time course of factors C, S, and S' (see text for explanation of factors). Bottom: use of the model to predict alertness at the end of a normal day (A) and at the end of a night shift following a day awake (B).

After rising at 07:00, subjects would, by 23:00, have been awake for 16 hours and be at the point shown by the arrow A, with an alertness score of 9. If they then worked a night shift, by 07:00 they would be on the '24h' curve, with an alertness score down to 5 (arrow B).

Importantly, it has been shown (by measuring behaviour and the electrical activity of the brain) that, if this score falls below a 'critical' value, the individual falls into a 'risk' zone where there would be a tendency to miss information relevant to the task in hand. This kind of finding has very important implications when hours of work, particularly those associated with night work, are considered.

## 3.2 Fatigue, body temperature, and sleep

Fatigue is a natural way of indicating that the body needs sleep. If a link exists between body temperature and the sensation of fatigue, then it is reasonable to enquire if the abilities to fall and to remain asleep are also related to body temperature.

Experiments have been performed to investigate how easy it is to fall asleep at different times of the day. One protocol consists of dividing the 24 hours of the day into $72 \times 20$-minute segments. In each of these segments, the volunteer is allowed 7 minutes to get to sleep in the comfort of a sleep laboratory, and a record of electrical activity in the brain (the electro-encephalogram, EEG) is kept which records the amount of sleep obtained. The subject is woken (if asleep) at the end of the 7 minutes, and required to remain awake for the next 13 minutes. The next 20-minute segment is treated in the same way, and the protocol continues this way until the end of the last segment.

The results (Fig. 3.2, top) indicate that getting to sleep is easier at night and more difficult during the daytime, which is what we would expect. In more detail, the ability to get to sleep mirrors the circadian rhythm of body temperature (see Fig. 1.2) — the higher the temperature, the less sleep is obtained in a 7-minute segment, and vice versa.

Other experiments have changed the pattern of sleep and activity and then considered how the length of time that we sleep, or the likelihood that, if asleep, we will wake up spontaneously in the next hour, depends upon the time of day. Again, the results have been compared with the rhythm of body temperature. Results (Fig. 3.2, bottom left) show that spontaneous waking is more likely to occur if the body temperature is rising or has reached a high value, but much less likely if the temperature is low or is falling.

Putting all these results together shows that, in the evening, when the body temperature is beginning to fall, there is an increase in fatigue, an increased ability to get to sleep, and, if already asleep, a decreased likelihood of waking up. This state of affairs continues while body temperature remains low (throughout the night). By contrast, in the morning, when body temperature begins to rise rapidly, we are most

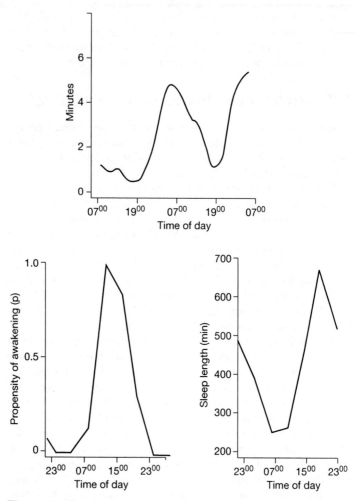

**Figure 3.2** Top: ease of getting to sleep. Bottom left: chance of a sleeping subject waking up in the next hour. Bottom right: number of minutes of sleep following different times of retiring.

likely to wake up and find it increasingly difficult to initiate sleep. These two factors together — getting to sleep and waking up — determine (Fig. 3.2, bottom right) how long we can stay asleep (at least, in a sleep laboratory that is always quiet, dark, and warm) at different times of the day. If we go to bed at the 'normal' time, we sleep about 8 hours — a result that accords with our own experience. However, if we fall asleep when the body temperature is about to begin to fall (about 18:00), there is the possibility of sleeping for 10 hours or more. By contrast, even if we manage to get to sleep at about 10:00 (and the rising temperature then will make this difficult), the sleep is likely to be shorter and broken.

There is one exception to this general parallelism between temperature, fatigue, and the ability to sleep. After lunch (at about 14:00) many of us feel tired and might take a short nap, even though body temperature does not normally fall much at this time. The phenomenon is called the 'post-lunch dip' — a reference to the dip in mental performance that occurs then. Its cause is uncertain, although it is thought to represent a shorter cycle or 'ultradian' rhythm (see Chapter 7). The post-lunch dip is also found even if we do not eat then, and a 'liquid lunch' containing alcohol certainly promotes it! It means that it is often possible to catch up on lost sleep by taking a nap at this time. It is, of course, the time of the siesta in many warmer parts of the world.

## 3.3 Sleep and sleep stages

As we know, sleep is not a uniform state. Sometimes we feel we have been deeply asleep, while at other times we recall that we have been dreaming. Watching the sleep of other people and of animals confirms that sleep can be quiet (with deep, slower breathing) or apparently eventful (with body movements and even sounds being made).

Recordings of the electrical activity of the brain, of eye movement, and of general muscle tone — from electrodes attached to the scalp, face, and neck — confirm that different types of sleep exist. These have been classified into three main groups. The first — rapid eye movement (REM) sleep, dream sleep, or paradoxical sleep — appears to be associated with dreaming. Our eyes, legs, and even our voice can act as though we are living the dream. The other main type of sleep — slow wave sleep (SWS) — is more difficult to wake us from and is, therefore, also known as 'deep sleep'. The third type of sleep is easier to rouse us from and is referred to as 'light sleep'.

During a normal night's sleep, we switch between SWS and REM sleep, getting through about four such cycles (REM/non-REM cycles) as the night passes. Each cycle contains progressively more light sleep, so that, by the end of the night, this is often the dominant form.

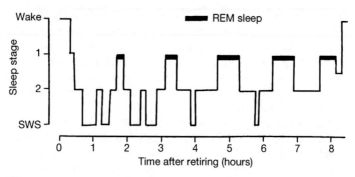

**Figure 3.3** The cycling between different sleep stages during the course of a typical night.

We generally awaken from light sleep — though not always, as the comments at the beginning of this section indicate. A typical profile of the sleep stages during the course of the night is shown in Fig. 3.3. This figure also shows that the first cycles are richest in SWS and that REM sleep is concentrated towards the end of the night.

This distribution of sleep stages can change, however. If we have recently lost sleep (after a week of trying to sleep in the daytime, for example) or have been awake for an unusually long time, then the amount of SWS in a REM/non-REM cycle increases, even though it is still concentrated in the earlier part of sleep. SWS appears to be associated mainly with the recuperative function of sleep, reflecting more the amount of prior wakefulness than the time when the sleep is taken. This means it has a very weak internal cause but a marked external one.

By contrast, the amount of REM sleep in a REM/non-REM cycle shows a strong internal cause. Experiments in which sleep is taken at different times have shown that the amount of REM sleep is very little affected by the amount of prior wakefulness but that it is affected instead by body temperature — the lower the temperature, the greater the amount of REM sleep. The fact that the temperature minimum is generally in the second half of sleep (see Fig. 1.2) explains why REM sleep is more concentrated in the later REM/non-REM cycles.

Even though the exact role of the different types of sleep is not known, sleep is certainly not a period when the body is merely marking time and ticking over idly. It must also be remembered that the almost universal habit of humans to take a single period of consolidated sleep, one lasting as long as a third of the total day, is comparatively rare in other species. Rodents, for example, have times of inactivity in the daytime, since they are nocturnal, but these are often broken by brief bouts of activity for foraging and cleaning.

Three main roles have been suggested for sleep in humans, but these overlap considerably. The first role is that sleep is a time when the activities of the brain change from those of acquiring, interpreting, and acting upon information obtained from the environment to those of consolidating daytime memories and experience. It would appear that REM sleep is important in this consolidation, though what the process of dreaming is telling us about this process is unclear — to physiologists, at any rate! Sleep is also a time when we conserve energy because it would be wasteful not to do so. Our faculties of sight, hearing, and smell do not equip us well for hunting at night, and trying to do so in our ancient past would probably have been something of a waste of effort. The lower body temperature associated with nocturnal sleep, coupled with the reduction in temperature caused by the body clock, enables us to reduce the amount of the body's energy stores that have to be broken down at this time of fasting (see also Chapter 6). A third role of sleep is that it provides an opportunity for the growth and repair of tissues, a function that seems to be associated with SWS.

Whatever might be the details of the roles of sleep, it seems that they are closely tied up with the secretion into the bloodstream of several hormones.

## 3.4 Sleep and the secretion of hormones

The secretory profiles of several hormones during a normal lifestyle (being asleep in the dark at night and awake and active in the light in the daytime) are shown in Fig. 3.4.

When a hormone is observed to be secreted into the bloodstream mainly during nocturnal sleep, the question arises if this time of maximal secretion is a function of the body clock rather than of sleep itself or of some other change associated with sleep (posture, lack of fluid intake, or darkness, for example). An answer to this question has been sought experimentally by requiring volunteers to sleep at abnormal times (during the daytime, for instance) and measuring the secretory profiles of the hormones during such a sleep. If sleep is the major influence, then the profile will change in accord with the amount by which sleep has been shifted, with high values still occurring during

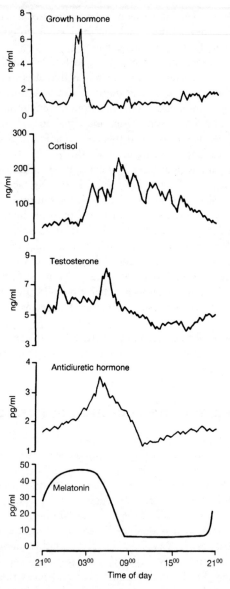

**Figure 3.4** The daily rhythms of several hormones measured in individuals living a normal lifestyle.

(daytime) sleep and low values during the time awake (at night). By contrast, if the body clock is the main influence, then the secretory profile will be little changed (because, as we shall discuss later, the body clock continues with little alteration to its timing). For some hormones, both the body clock and sleep exert influences of comparable size, as a result of which the secretory profile will show two smaller peaks, one at night and the other during the sleep. Investigations of the effects of posture, lack of fluid intake, and darkness are performed in a similar way, the factor under consideration being moved from the middle of the night to the middle of the day, while other factors are kept unchanged.

The secretion of growth hormone — a hormone that regulates many aspects of our metabolism — is dominated by the time of sleep rather than the body clock, and is linked to the electrical changes in the brain that are associated with SWS. This strongly suggests that the secretion of growth hormone is associated with the role of sleep in growth and repair, and the metabolism of food stores during fasting (see Chapter 6). It is interesting to note that severe physical exercise in the daytime is sometimes associated with increased SWS and growth hormone secretion the following night; this would support the idea of recuperation via both SWS and the secretion of growth hormone.

By contrast, the hormones cortisol and melatonin seem to be influenced far more by the body clock than by sleep. Cortisol secretion, which peaks at the end of a normal sleep, is also associated with the metabolism of food stores during a period of fasting, but it also exerts an immunosuppressive role (reducing our immune responses), and acts as a general 'anti-stress' hormone. The role of cortisol in immune responses will be covered in Chapter 14.

Melatonin is a hormone that is of particular interest to those who study biological rhythms. It is produced by the pineal gland and is related chemically to serotonin, a substance whose release is associated with sleep. The pineal gland is situated deep inside the brain in humans, but it is much nearer the surface in species in which the cerebral hemispheres have not enlarged as much as in humans. Indeed, in the Tuatara lizard, to be found in New Zealand, the gland is on the surface of the brain and directly picks up light signals, acting as a 'third eye'. In many vertebrates, the pineal gland seems to have retained a function associated with light in the environment, because it is important, via its secretion of melatonin, in seasonal behaviour such as breeding cycles.

In humans, melatonin seems to be more important when daily rhythms are considered. It has been called the 'hormone of darkness', not only because it is normally secreted only at night but also because its secretion is suppressed by bright light. Moreover, there are receptors for melatonin on the cells of the body clock, and there is evidence that melatonin can adjust the body clock. For this reason, melatonin is also known as an 'internal zeitgeber'. A further effect of melatonin is to reduce body temperature. It does this by causing blood vessels in the skin to dilate, so increasing the supply of warm blood to the surface of the body and allowing this heat to escape. Note that the fall of body temperature in the evening, necessary to increase our sense of fatigue and to facilitate getting to sleep, is produced by a combination of the body clock, an environment conducive to sleep, and by the loss of heat through the skin produced by melatonin secretion.

The antidiuretic hormone (ADH) is responsible for regulating the rate at which water is lost from the body via the kidneys. On a hot day, when we tend to be dehydrated and so wish to conserve body water, ADH is secreted into the bloodstream in greater amounts; after a daytime drink of water, its secretion is suppressed, and urine flow increases. Secretion of ADH at night is increased both by the body clock and by sleep. As a result, our drink at supper time is less likely to wake us up due to the need to empty our bladder. This is all the more important since lying down increases the return of blood back to the heart and this increases urine flow. In order to appreciate the effect of an increased secretion of ADH at night, try lying down for 8 hours after drinking a pint of water in the daytime, when there is a fall rather than a rise in ADH secretion!

Table 3.1 summarizes some of the findings discussed so far. The hormones adrenaline and noradrenaline, secreted mainly in the daytime during times of activity, will be considered in Chapter 4.

As Table 3.1 indicates, if sleep is taken at an unusual time, changes in the release of a hormone or the type of sleep will depend on the relative importance of external factors and the body clock for that variable. Phenomena that are influenced mainly by the body clock will be affected more (you are sleeping at the 'wrong time' as determined by this clock) than those determined mainly by external factors. Whether such abnormalities lead to sleep being less 'refreshing' is not known. However, for those whose time of sleep is often changed (air crew involved in long-haul flights and night workers, for example), the knowledge that the profile of hormone secretion during sleep is often abnormal must give cause for concern — or unease at the very least.

**Table 3.1** A comparison of the profiles of some hormones and types of sleep during night-time and daytime sleep

|  | **Night-time sleep** | **Daytime sleep** | **Major influence** |
| --- | --- | --- | --- |
| Growth hormone | Mainly first part | Mainly first part | Sleep |
| Cortisol | Rises throughout sleep | Falls throughout sleep | Body clock |
| Melatonin | High throughout | Low throughout | Body clock |
| ADH | High throughout | Quite high throughout | Both |
| SWS | Mainly first part | Mainly first part | Sleep |
| REM sleep | Mainly towards end | Mainly at the beginning* | Body clock |

*REM sleep can 'squeeze' SWS from the first part of sleep towards the middle of it.

# Now try this: measure your own body temperature rhythm and ratings of alertness

## Measuring temperature and alertness

A whole range of thermometers (for measuring oral, forehead, or eardrum temperatures, for example) is now available. Of these, the measurement of oral temperature (sublingual, under the tongue) with a clinical thermometer (one that records the highest temperature reached) is most common. But, in order to get reliable readings, it is important to observe the following precautions:

- Keep your mouth closed and breathe through your nose for the 5 minutes during which the temperature is being taken. Note that, for an accurate reading, the thermometer needs to be in place for *at least* 5 minutes.

- Make sure that you have not been talking, drinking, or eating for the previous 10 minutes.

If you observe these precautions then sublingual temperature is a reasonably reliable measure of **changes** in body temperature. However, oral temperature is consistently lower than other measures of body temperature — oesophageal, gut, or rectal temperature, for example — and it is these methods that are preferred in a research setting where the **absolute** value for body temperature is required.

Measuring alertness is subjective, and a simple way to do this is to record an 'alertness score' on a 7-point scale. An example of this is as follows:

| Alertness score | Meaning |
|---|---|
| 7 | Fully alert, not at all fatigued |
| 6 | Very slight fall in alertness, very slightly fatigued |
| 5 | Slight fall in alertness, slightly fatigued |
| 4 | Moderate fall in alertness, moderately fatigued |
| 3 | Fairly marked fall in alertness, fairly marked fatigue |
| 2 | Very marked fall in alertness, very marked fatigue |
| 1 | Extremely marked fall in alertness, extremely marked fatigue |

Clearly, you might wish to define the scores differently; that does not matter, as long as you always use the same definitions. Also, different individuals will score themselves differently under any set of conditions. Again, this does not matter, provided they are consistent. The important thing is to measure the **changes** that take place.

## Obtaining 'resting curves'

The problem now is to deal with the effects of the external causes on the body temperature and fatigue rhythms. A starting point would be to produce 'resting curves'. To do this, the body temperature should not be taken until the subject has been sitting quietly in a comfortable environment indoors for about 30 minutes. The alertness score should be recorded at the same time, but before the temperature is known. It is also much better if the previous alertness scores are **not** known, since this knowledge might bias the new value. (This problem does not arise with a more objective measure such as body temperature, of course.)

Measurements could then be taken every 2 hours while awake, for example. It would also be valuable to record the alertness score and body temperature in the morning (but while still in bed), and also after having relaxed in bed, just before turning out the light at the end of the day.

Using this method — one in which external factors and the conditions of measurement have been standardized as much as possible — should enable the rising values in the morning and the falling values in the evening to be demonstrated; that is, the results should reflect the internal cause of the temperature rhythm.

However, the value for body temperature obtained immediately after waking in the morning reflects partly the fall of temperature that accompanies sleep (probably due to the change in posture associated with it). This will contribute to the low value for alertness measured at this time. However, alertness is low also due to not having woken up properly. This state is known as 'sleep inertia' and can be a problem if important tasks are performed too soon after waking up (at whatever time this might be). This factor is incorporated into more complex versions of the model that is shown in Fig. 3.1.

## Obtaining values at night

To get values during the night requires waking up regularly, and this makes things less fun and more of a chore. There is also the problem that these values have not been obtained under the same conditions as those during the daytime: they reflect the fact that one has been lying down and asleep (see above). One way to get the values without these effects, therefore, would be to stay up all night and continue to take readings every 2 hours and under the same conditions as during the daytime. However, when the effects of sleep loss begin to show, it becomes a problem to decide how 'normal' such values are, particularly those of alertness.

Comparing nocturnal values obtained while having remained awake with those obtained after having woken from sleep would give you a measure of the effect that sleep and lying down had caused. Temperatures after sleep would be lower (compare the two curves in Fig. 1.4, top) and the effect of sleep inertia upon the alertness scores would be large.

## Effects of naps and activity

Does a sleep in the daytime also cause a fall in alertness due to sleep inertia and a fall in body temperature, compared with values obtained when sitting quietly? To find this out, take a nap, and then take your measurements on waking up from this. Compare the measurements with those obtained **at the same time of day** (to take into account the effects of the rhythm due to the body clock) when you have stayed awake and been sitting quietly.

To investigate the effects of different amounts of activity (standing, walking, and so on), carry out the activity for 30 minutes before taking your measurements. Always compare the values with those taken after having been sitting down at the same time of day, in order to correct for the rhythm due to the body clock.

You should find that the body temperature and alertness scores at any time of day are lower as you pass along the sequence — exercise/walking/sitting, doing something/sitting, relaxing/lying down, awake/asleep. Therefore, we need a 'standardized' circumstance under which to measure body temperature and alertness. This is, as the above protocols suggest, sitting down quietly.

One implication of such findings is that fatigue can be lessened by bouts of exercise. How practical this might be (for a lorry driver or machine operator, for example) is another matter, but at least there is the possibility that we can sometimes take steps to oppose such feelings. Unfortunately, however, the effects appear to last for a short time only.

## The collection of large amounts of body temperature data

Such experiments are time-consuming and require considerable care by the subject. In scientific studies, body temperature has been measured automatically every 6 minutes or so, the measurements stored on a microchip and then downloaded to a computer for analysis. Measurement of rectal temperature is the preferred method (at least by the experimenter!), since it gives a more accurate measure of the body's core, but some subjects find this unacceptable, particularly if they are living normally rather than in a laboratory setting. Sublingual temperature is too demanding of a subject (due to the effects of talking and drinks upon mouth temperature), and this method, together with measurements of eardrum temperature, are impossible when the subject is asleep. Moreover, frequent sampling is inconvenient. A 'temperature pill' can be swallowed and used to record body temperature as it moves through the gut, but it is expensive. Temperature in the armpit, or axilla, has also been used as an alternative, but it is not reliable, particularly if the subject goes outdoors into a cool environment.

Alternatives to alertness scores will be considered when mental performance in general is discussed (see Chapter 5).

# Chapter 4

## Your body clock, the waking period, and physical activity

We have already considered how the body clock and melatonin secretion, by lowering body temperature in the evening and at night, increase fatigue and decrease alertness, and so help to promote and maintain sleep. This is a time of recuperation, growth, and conservation of energy. In the daytime, the opposite effects take place; we are now alert and ready to be active. Whatever form these activities take, they will require energy. That energy comes from the breakdown of food and is used both for physical and mental activities. These issues will be the topics of this and the next two chapters. Here we consider the cardiovascular and respiratory systems, the provision of energy by metabolism, and physical performance.

## 4.1 Adrenaline and noradrenaline — the hormones of action

Adrenaline and noradrenaline are produced by the inner part of the adrenal glands (structures above the kidneys). Noradrenaline is also released by the nerves of the sympathetic nervous system. Both show a circadian rhythm with highest values about midday and lowest values during nocturnal sleep. Their secretion is parallel to the 24-hour rhythm of alertness.

Adrenaline is conventionally regarded as a 'stress' hormone. This means that it is released when we are angry or anxious, and its release plays a major part in the 'fight and flight' reaction in us and in other animals. At the same time, there is an increased activity of the sympathetic nervous system, which performs a function very similar to that of adrenaline. Adrenaline exerts many effects upon the body in order for the fight and flight response to be effective. It makes the heart pump blood round the body faster (we have all felt the pounding and racing of our hearts when we are angry), and it dilates the airways leading to our lungs so that we can breathe faster. It also causes an increase in the release of energy from body stores. Thus, it causes muscle glycogen (the form in which glucose is stored in muscles) to be broken down for increased muscle effort, and for fat (from fat storage depots) to be broken down for use by the rest of the body. In these ways, adrenaline enables the body to be best able to deal with danger and to be most efficient physically. The fact that stressful circumstances, such as anger or danger, are not conducive to going to sleep is largely due to adrenaline.

When there is no undue stress (as in the case of people living normal lives), adrenaline is still released in the daytime, but in smaller amounts, and the sympathetic nervous system is active, but less so. In these circumstances, adrenaline and noradrenaline are still important, since they 'tone up' the body so that it can perform more efficiently. They are part of a mechanism by which we feel alert rather than fatigued during the daytime, and are better able to perform physical and mental activities during this time. Equally important are the decreases of adrenaline and noradrenaline secretion in the evening, when the body prepares for sleep.

In summary, the daily rhythms in adrenaline and noradrenaline, and the sympathetic nervous system rhythms are like that of body temperature with regard to their general timing, their function, and their integration with the sleep–wake cycle.

Some of the effects of adrenaline are illustrated by the study (considered in Chapter 2, Fig. 2.1) upon unstressed army volunteers who stayed awake for 72 hours in constant conditions and ate regular meals. In addition to 'target practice' and recording fatigue, the volunteers gave urine samples, which were analysed to estimate the rate of excretion of adrenaline. (This value reflects the concentration of this hormone in the blood, and collecting regular urine samples is considerably more convenient than taking regular blood samples by venepuncture.) Figure 4.1 shows again the fatigue ratings and the speed of shooting at a target, but also includes the rate of excretion of adrenaline. Note that the rhythm of adrenaline excretion in urine (equivalent to the rhythm of concentration of this hormone in the blood) shows peaks and troughs that are timed contrary to those shown by fatigue, but timed the same as the peaks and troughs shown in the speed of shooting at a target.

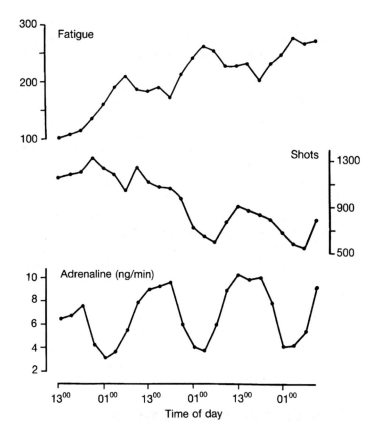

**Figure 4.1** 3-hourly ratings of fatigue (arbitrary units), speed of firing at a target (high score indicates high speed), and urinary excretion of adrenaline in a group of soldiers staying awake for 3 consecutive days.

## 4.2 The demands of exercise

Can the body perform physical work equally well at all times of the 24 hours? Consider the biochemical and physiological changes that are required to exercise efficiently.

- The basic biochemical process is one by which glycogen stores in the muscle and fat depots are broken down to release energy which can then be used to drive the muscles.

- An efficient breakdown of glucose and fat requires oxygen. An increased blood supply to the exercising muscle brings this extra oxygen, and also removes the waste products of metabolism (carbon dioxide and lactate).

- The increased blood supply to the muscles requires not only increased pumping activity by the heart (brought about by increases in the rate and force of contraction) but also a preferential distribution of this extra rate of supply to the active muscles. This extra delivery of blood must not be at the expense of a sufficient supply continuing to the brain and other vital organs. In order to facilitate blood supply to the active muscles, the blood flowing to certain organs (for example, viscera and kidneys) is reduced.

- An increased rate and depth of breathing ensures that more oxygen is taken into the lungs, from which it passes into the blood, and that more carbon dioxide can be removed from the blood and expired into the atmosphere. The amount of work that the respiratory muscles must do is reduced if airway resistance decreases; that is, the airways are dilated.

This list does not include the later changes that are necessary to prevent the build-up of excessive amounts of heat from the exercising muscles, but it does indicate how many processes are involved. Adrenaline, noradrenaline, and the sympathetic nervous system are involved in all of them.

## 4.3 Rhythms in maximal physical performance

When the performance of elite athletes in competition is considered, it is found that best performances and the breaking of world records occur mainly in the early evening. This has been attributed in part to the fact that this time is close to the time when, in a resting subject, the rhythms of the cardiovascular and respiratory systems, of adrenaline and noradrenaline secretion, and of body temperature are all at their peak. However, other factors such as media coverage and the presence of crowds might well be at least as important.

Nevertheless, under more controlled conditions (in a laboratory setting, for example), daily rhythms of performance persist, and a selection of these is shown in Fig. 4.2. The examples show, from the top down, the maximal power output on a swimming ergometer, the

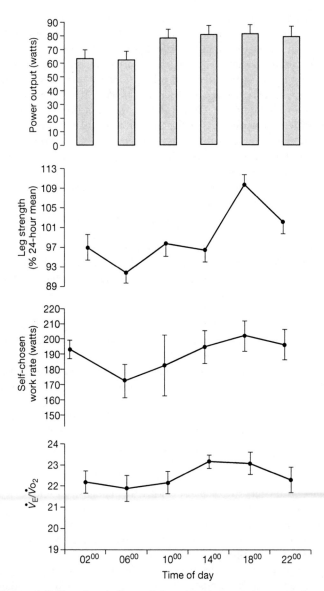

**Figure 4.2** Circadian rhythms of, from top downwards, maximal power output in swimming, maximal force development by the leg, self-chosen work rate, and the ratio of ventilation to oxygen uptake.

maximal force developed by the leg muscles, the self-chosen work rate on a cycle ergometer, and the ratio of ventilation to oxygen uptake. These will be referred to individually later.

The rhythms that are most important to an individual depend on the type of activity that is involved. Compare, for example, the requirements of a gymnast, sprinter, and weight lifter, or the requirements in the different disciplines in the triathlon, pentathlon, heptathlon, or decathlon. In the case of endurance events, such as long-distance time trials for cyclists, the marathon, or certain cross-country skiing events, another important issue is that of removing sufficient heat from the body.

Since resting body temperature is higher in the afternoon and evening, the rise in temperature produced by exercise means that, in endurance competitions at this time, body temperatures are likely to exceed the optimal value. With morning competition, there is the advantage, since the body temperature starts from a lower value, that more exercise can be performed before the risk of 'heat stress' arises. For this reason, starting these events in the morning, when the body as well as environmental temperatures are lower, has benefits for the participants. The argument — that the lower body temperature at the start of an event held at this time will compromise performance overall — appears to be incorrect when sustained exercise is considered. This is because, even though individuals in the morning do, indeed, start more slowly than those starting in the afternoon, this slow start later becomes an advantage, since it also reduces the build-up of lactate in the early part of the event.

Lactate is produced when oxygen demand by the muscles outstrips supply. It is associated with muscle fatigue. This is reflected in a reduced performance, and the muscles break down their store of glycogen more rapidly. The so-called 'lactate threshold' indicates the exercise intensity at which the concentration of blood lactate begins to rise disproportionately in comparison with the amount of exercise. It is thought to restrict the upper limit for sustained exercise. An endurance athlete performing in the morning is likely to be operating well below this lactate threshold, thereby conserving glycogen stores for later in the competition.

Figure 4.3 illustrates these arguments. In the afternoon, the higher resting temperature means that the chosen work rate is initially higher than with morning exercise, but it cannot be sustained, due to rising lactate levels and the increased likelihood of heat stress.

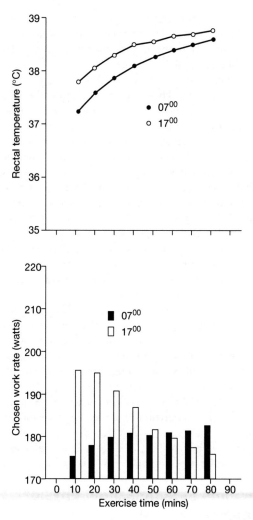

**Figure 4.3** The changes in temperature (top) and self-chosen work rate (bottom) with sustained exercise in the morning and late afternoon.

For events that last for several minutes, it is the oxygen supply to the muscles that is most important. Even though studies of maximal heart rate and rates of ventilation and oxygen uptake do not show convincing circadian rhythms, rhythms in performance are present.

In particular, maximal strength and maximal power output display a clear circadian rhythm, as do performances in running and swimming events, all of them showing best performances in the daytime and worst, at night (see Fig. 4.2, top trace).

Two other factors seem to be involved — muscle performance itself and perceived exertion. These factors will apply also to sprints, to 'explosive' activities (jumping, for example), and to tests of muscle strength (weight lifting, for example).

One approach to investigating these daily rhythms in physical performance in humans is to consider the individual components that constitute a particular sporting activity. Many laboratory-based studies indicate that various muscle groups — involved in a range of activities including grip strength, back strength, squat jumps, leg extension, and leg flexion — show rhythms that are in phase with those of resting body temperature (see Fig. 4.2, second trace down).

It is generally believed that this coincidence reflects the molecular processes that take place when a muscle contracts. Since muscle contraction is a biochemical process, it will increase with increasing temperature. This is one of the reasons why, particularly in cold weather, athletes take care to 'warm up' properly. The existence also of a daily rhythm of body temperature means that the joints and muscles of the body will be cooler in the morning, and so there will be a daily rhythm of joint stiffness and body flexibility. Therefore, gentle stretching exercises are usually recommended prior to exercise in the morning, in order to reduce the risk of injury.

In summary, physical activity is most efficient in the late afternoon. The amplitudes of these rhythms of athletic performance tend not to be very large, a matter of only a few percent in some cases. Even so, for elite athletes, the difference between winning a race or setting a new record and a mediocre performance can be only fractional.

## 4.4 Rhythms in submaximal physical performance

As we are not all elite sportspersons, the question arises as to how relevant the above findings are to most of us? Also, athletes are not always involved in competition, much of their time being spent in training. Are daily rhythms relevant to them in these circumstances?

If individuals are asked to do the same physical task at different times of the 24 hours (for example, pedalling on a cycle ergometer), they can also be asked to record how much of an effort it appears to be. It is commonly found that the task — even though it was equally demanding physiologically and biochemically — is felt to be most difficult in the middle of the night. The experiment has been repeated in a different way, by asking the subjects to exert themselves equally at different times of the day, and then assessing how much work they actually choose to do (but without letting them know this amount). At night, subjects choose to do less work than during the daytime (see Fig. 4.2, third trace down). In both cases, there is a parallelism between the rhythm of resting body temperature and muscle performance.

The daily rhythms of oxygen consumption and ventilation that are evident while resting are observed also during light or moderate exercise (although they tend to disappear when exercise is sufficient to elicit maximal oxygen uptake). The rhythm of ventilation has a higher amplitude than that of oxygen consumption. This is probably due to an increased constriction of the airways at night and early in the morning (in turn due to less adrenaline in the bloodstream and less sympathetic nervous system activity at these times). As a result, the ratio of ventilation to oxygen uptake shows a daily rhythm which, once again, is in phase with the temperature rhythm (see Fig. 4.2, bottom trace). This means that exercise in the early morning is less easily tolerated than later in the day because, for any oxygen requirement by the body, increasing ventilation to get this amount of oxygen into the body is more difficult. This suggests also that there is a link between increased breathing difficulties and perceived exertion.

Figure 4.4 is a record of the mean heart rate and percentage of time that could be described as 'active' in a group of football players taking part in a sponsored indoor game that lasted 91 hours. During this time, they were allowed neither sleep nor rests (apart from a 5-minute locker break every hour). It is instructive to compare Fig. 4.4 with Fig. 4.1.

In both cases, there is a progressive decline in performance at the allotted task, and this is coupled with evidence of a parallel fall in general activity, as assessed by heart rate. This can be called 'fatigue', and will have both mental and physical components for the football players. Superimposed upon this decline are the parallel circadian rhythms of heart rate and the percentage of time spent being 'active'. The troughs are at night and the peaks during the daytime.

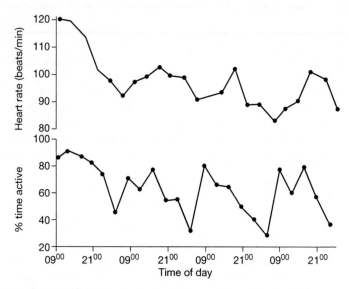

**Figure 4.4** The heart rate (top) and percentage of time 'active' (bottom) during a sponsored football game lasting 91 hours.

The activity required of the players was never maximal. Even so, it is almost certain that the perceived exertion required during the troughs of activity did not fall as much as did the amount of activity itself. In other words, in the middle of the night, the players were doing much less work but their feeling of doing less had not fallen by the same proportion. These findings mean that the same amount of training exercise will be perceived as being harder at some times than at others.

There are even wider implications. For individuals whose occupation is 'physical' rather than sedentary, they function at about 40% of their physical capacity when at work. If this work is carried out at night, or if it is done after sleep loss, it will be perceived as being physically harder. There are implications also for those, particularly sportspersons who wish to train, who have yet to adjust their body clock to a new time zone after intercontinental flights.

It would be very unfair to suggest that sportspersons need only consider 'brute force', and that rhythms of mental processes are unimportant to them. This is not the case, and this is the topic of the next chapter.

# Now try this: making further use of our rhythms

We have now reached a position where people living normal routines can benefit from their knowledge of body clocks. Below are three possibilities.

## 1. A comfortable room temperature

How comfortable we feel is determined not only by the temperature of the body (more accurately, the temperature of the blood going through the brain) but also by that of the skin. In order to feel comfortable we must get the balance right. Dressed conventionally, we do this by controlling both body temperature and ambient temperature at 'normal' values (about 37°C and 20°C, respectively). If either is higher or lower, we feel uncomfortable.

Nevertheless, we can make ourselves feel more comfortable even if body temperature is higher or lower than this 'normal' value. For example, if we have raised our body temperature by exercise, then taking a cold shower, a swim, or a cold drink makes us feel more comfortable. Likewise, if our body temperature is low, warming our hands in front of the fire is very comforting. A similar line of reasoning — warm the skin if the body temperature is cold — explains the ideas behind 'one for the road' and the 'stirrup cup'. The nip of alcohol causes the blood vessels to dilate and so warm the skin. This makes us feel comfortable in spite of the tendency for body temperature to fall. As a final example, consider the passage of air across the skin of our face. If our body is becoming too warm (through sunbathing, for example), then the air is a 'pleasant breeze'; but if our body temperature is tending to fall (sitting down in wintertime), then the same air movement is a 'draught'.

In all cases, the feeling of comfort is immediate, and is not associated with a change in the temperature of the blood flowing through the brain. It arises because a high body temperature can be offset by a low skin temperature (and vice versa) to produce the sensation of 'comfort'. However, this feeling of comfort might give a false sense of security since the underlying problem (the wrong blood temperature) is not being corrected; indeed, in the case of a nip of alcohol and cold weather, the dilatation of skin blood vessels actually worsens the problem.

We can now look at daily changes in body temperature. In the evening, body temperature is being reduced and so tends to be slightly higher than required. You can help this fall in body temperature by being inactive (by sitting in front of the television or, perhaps, reading this book!). If you are quite active physically then you can maintain a sense of comfort if you turn down your room thermostat a few degrees in the evening, or if you roll up your sleeves and aid heat loss this way. If you do neither of these things, then you might find that the room seems 'stuffy'.

In the morning, body temperature is being raised and so it tends to be slightly lower than required. This calls for warmer skin and environmental temperatures in order for us to feel comfortable. Cheaper alternatives to turning up the central heating are to do a brief bout of **gentle** exercise (see below), which increases body heat from the inside, or to wrap up well, which reduces the loss of heat to our surroundings.

## 2. Exercise and physical exertion

The usefulness of gentle exercise in the morning to help warm us up has just been mentioned; it will also help to increase our alertness. Further, exercise at this time will promote the loss of fluid from the discs between the vertebrae (see Chapter 1). However, remember that the slightly swollen intervertebral discs and the increased joint stiffness might increase the risk of injury. Be careful, therefore, when exercising and, particularly if you are lifting heavy loads, make sure that you are doing so correctly.

For more severe exercise, the best part of the day is late afternoon because the body is most efficient then and you will be able to push yourself harder. Leaving exercise too late in the day might raise body temperature sufficiently to make it harder to get to sleep afterwards. Heavy exercise very soon after waking is not such a good idea; the rise in heart rate and blood pressure caused by the exercise will be added to those that occur naturally as we pass from the sleeping to the waking state. Excessive stresses might damage the heart and the blood vessels of vulnerable individuals, and it might not be a coincidence that cardiac disorders, including exercise-induced 'angina', are most frequent at about, or in the hours immediately after, the time of waking (see Chapter 14).

We appreciate that the best times for exercise are often inconvenient to individuals who wish to train but also have a regular daytime occupation. For them, the only times available seem to be before work or after it in the evening. In such cases, is it possible to have a lunchtime work-out? This could have the advantage of recharging motivation for work in the afternoon — providing that the session is not too strenuous, in which case it might prove to be counterproductive!

## 3. Waking up and getting to sleep

The advice that was given in Chapter 1, for changing the habits of 'larks' and 'owls', can be used more generally to promote waking up or getting to sleep.

Waking up fully can be promoted by gentle exercise, as well as by carrying out tasks that require some physical effort — as long as they are not too demanding (light housework or weeding the garden, for example; but definitely **not** digging). This way, the effects of the body clock in raising body temperature and adrenaline and noradrenaline release are being accentuated. Also, a stimulating environment, such as one with lively music and bright lighting (in natural lighting, if possible) can help.

In the evening the aim is to prepare the body for sleep. Hot or cold showers, arguments, and heavy exercise are not advised! Many people practice relaxation techniques, take a soak in a bath of warm (neither hot nor cold) water, make a bedtime drink, or read quietly in bed before turning out the light. All these things promote the decreases of body temperature and adrenaline and noradrenaline release that are taking place naturally. A quiet and dimly lit environment can help, with low levels of lighting promoting the release of melatonin (with its temperature-lowering effect) into the bloodstream.

# Chapter 5

# Your body clock and mental activity

At night, physical work seems harder to do than normal, and it is done less well. Does the same apply to 'mental performance'? Here we consider mental performance to include using our senses to take in information about the external world (sensory input), understanding events in our external environment (cognitive performance), decision making and other intellectual processes, and producing coordinated muscular activity to achieve a particular task (motor performance).

We are likely to be concerned if we perform badly at some times, and the effects of poor performance will also be of interest to managers and the general public. Managers will wish to know if the body clock means that there are times when the output of their workforce is below standard; and the general public will be concerned about the possibility that safety in industrial processes, public services, transportation, and so on will be adversely affected at some times.

## 5.1 Some problems with measuring performance at work

As soon as an attempt is made to measure 'performance' at work, a whole series of problems arises. How would you assess the performance of a plumber, a lorry driver, a teacher, or a magistrate? For the plumber or magistrate, is it the numbers of jobs or cases completed successfully? However, some might be more difficult than others. For the teacher, is it how well the class has understood the lesson? However, the topic might be difficult, or the pupils not very bright. How do you judge the lorry driver — by whether or not he is late or by how many times he has lost his way? Surely, both depend upon the traffic conditions and the difficulty of the route.

Sometimes, 'performance' is considered from a slightly different viewpoint, and only poor performance — as assessed by accidents or errors — is taken into account. In some cases, even this can be complicated. How would you establish reliably if a magistrate, politician, inspector, or teacher had made an error? Moreover, even if errors could be assessed satisfactorily, there need not be any clear link between a full-blown error or an accident and a poor level of performance — as any self-critical driver knows.

Even then, it is difficult to be sure to what extent more errors or a poorer performance are due to the body clock. Consider traffic accidents. If it were found that more accidents occurred in the winter months between 18:00 and 20:00 compared with 12:00 and 14:00, is this because, during the early evening, lighting is worse, there is more traffic, or drivers have been working longer and are more tired? Or is it simply that weather conditions are worse?

It might be thought that performance at many factory jobs would be easier to assess since there is an end product that can be measured in terms of quality and quantity. However, if the workers perform better at one shift (and 'better' might refer to the quality or quantity of the product), is it because the better shift has more conscientious workers, better working conditions, less distractions, better supervision, or a conveyor belt that moves faster? Or do the workers possess a body clock that enables them to work better at certain times of the day?

If we want to know the role that the body clock plays in our mental performance, these other, external influences must be controlled or eliminated.

## 5.2 Some improvements — but further problems

To remove or control the above problems, the following conditions need to be observed.

1. The same workers are investigated at different times of the day.

2. The workers have been on duty for the same amount of time whenever measurements are made. (This is because it is known that performance deteriorates after several hours of work.)

3. Ambient factors (lighting, heating, noise, traffic conditions, weather) are the same.

4. The work being investigated is always of the same type and difficulty.

5. The individual, not a supervisor or the speed of a conveyor belt, decides the rate at which the task is carried out (This is termed a 'self-paced task'.)

Condition (1) can be fulfilled by requiring the same workforce to work all shifts in a rotating shift system. With sufficient manipulation of the rota, it would be possible to arrange for the workforce to have been working for the same number of hours at different times of the day — condition (2). The conditions described in (3) are sometimes fairly easy to control (for example, the amount of noise and environmental temperature), but others (traffic density or natural lighting) can be more difficult or impossible. In some jobs, condition (4) can be fulfilled as it is possible for similar types of work to be tackled throughout the 24 hours, but this is not always so. For example, teaching at night is unlikely to prove popular (with teacher or pupil!) and the night shift in a hospital ward cannot be made equivalent to the day shift (except for intensive care units which are continuously busy). Finally, condition (5) cannot be fulfilled in all occupations. For ambulance operatives and air traffic controllers — those in 'supply-and-demand' situations — the demand is due to outside influences, and the workers have to respond; their workload is not self-paced.

## 5.3 Errors measured in the workplace

It must be apparent by now that fully satisfactory studies of rhythms in mental performance in the workplace are likely to be rare. Some studies do exist, however, and the results from a selection of these are shown in Fig. 5.1.

We do not need to discuss the results in detail, but make the following general points:

- Errors tend to be more frequent at night than during the day.

- Errors increase temporarily immediately after lunchtime.

The value of such 'on site' studies cannot be exaggerated, and the care with which they have been performed cannot be overestimated. They provide the best estimates by far of the role played by the body clock in 'real' tasks. They also serve two further roles. First, they act as a standard against which the results of laboratory-based tests

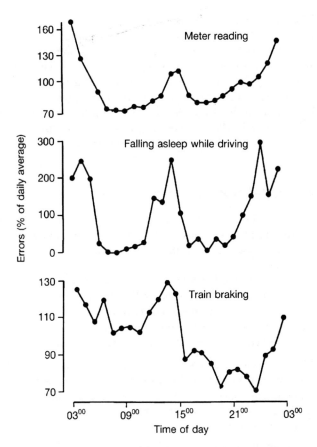

**Figure 5.1** Relationship between time of day and frequency of errors at three tasks measured on site.

(see below) can be measured. Second, they confirm the difficulties of interpretation that can arise.

To take this second point first, consider the frequency of errors made by Swedish train drivers (Fig. 5.1, bottom). ('Error' here refers to the failure of the driver to respond to a warning light that comes on when a signal at red is passed.) If the brake is not applied by the driver, then it is applied automatically; it is the frequency of these automatic braking incidents that is recorded as an error. From what we already know about the body clock and a consideration of the 'post-lunch dip', the high frequency of errors during the night and after lunch are as expected; but why are errors so infrequent in the evening? It is because the warning light is more easily visible in the evening twilight than in the brighter light of the daytime. This external factor offsets the fact that performance in the evening due to the body clock might be deteriorating. It will also be noted that, even though the warning light is most easy to see at night, this does not completely offset the decrement in performance caused by the body clock at this time.

## 5.4 Mental performance tests

By now, it might be wondered whether there is any satisfactory way to measure mental performance. The many complications seem to preclude this possibility, even though the importance of success is as great as ever.

One approach that has been adopted by psychologists has been to break down 'real' tasks into a series of 'simple' components, and then to devise tests for assessing these simpler components. It can be argued that these tests are a reliable indicator of performance when tested under controlled conditions. The simple components are as follows:

- **The sensory component.** Information has first to be taken from the environment and passed to the brain. The eyes are generally important, but the senses of touch, hearing, taste, and smell can all also be used. (Consider a chef, carpenter, piano tuner, tea taster, or perfumier, for instance.)

- **The central component.** This is the processing of inputted information by the brain. It will include reasoning or thinking, and might make considerable use of long-term memory and experience.

- **The motor component.** This refers to a type of action. It might be as straightforward as writing down or calling out an answer, pressing a button, and so on, or it might be more complex and require manipulative skills, such as wiring an electrical component or threading a needle. It might also involve coordination between several movements, as in driving a car.

There are two other components that tasks can possess — namely, short-term memory and vigilance. Like the central component, they involve processing information within the brain, but they are slightly different from what is normally meant by thinking or drawing on memory and experience.

- **Short-term memory.** This is the phenomenon by which we can remember a telephone number long enough after looking it up to be able to dial it. It also enables us to remember the cards that have been played in a hand of bridge, the objects on a tray in the game of Pelmanism, or reading a meter and recording the result (see Fig. 5.1, top). Short-term memory is a temporary store of information and often it is irrelevant to convert that information into a long-term memory (a type of memory that enables us to recall events hours, days, or even years later). Thus, although it might be useful to remember a particular telephone number (and so convert a short-term memory into a long-term one), this is often not the case — we are unlikely to want to remember every round of cards that we have played! Short-term memories are often required to be forgotten, therefore. A few people are able to remember everything ('photographic memories'), but this could lead to remembering a mass of unwanted material. Thus, things have to be actively 'un-remembered'.

- **Vigilance.** Through vigilance, we detect changes in our environment, changes in the traffic when driving, a new 'blip' on the radar screen, or a faulty product on a conveyor belt. One characteristic of a task requiring vigilance is that it can span a long period of time, minutes or hours, or even a whole work shift.

## 5.5 A selection of simple tests

To illustrate the assessment of these components of performance, we will now consider a small selection of the tests that have been devised. In the 'Now try this' section at the end of the chapter we suggest that you use these kinds of test to measure your own performance.

In most cases, the subject is asked to perform the test 'as quickly and as accurately as possible'. The test is scored two ways — the time taken to complete it (or the number of examples attempted in a set amount of time), and the number of mistakes that have been made.

### Group 1: tests which stress the sensory input

These tasks require a simple response and not much thought, coupled with the need to sift through large amounts of data in order to find a 'target' amongst unwanted information (see Fig. 5.2).

```
AGBHJTYOWLNMPASDDQGTUITLMDJEX
RYHSKMXHWYUSGHWWMLQWMSPLEUJDM
UJKSEHMSKQXGKLLQTHOMPDHJJEKLO
YWTFGHIKMCCOWLXPMTXWKIRLMHJDB
POEKJDNMFKURHHNRJKSSLPLQXYJDD
QWBDHJKPLMCBSXAZIDZKLEGHLCVVE
PDMSMBBOJWSQMNXLUDTTKWKLFMRQL
BNKLFMNSGBEOPAGVSBHKFMNXBVSJH
ILDKJMSVSKIRGQNBXTYUOLLKHWTPS
WJKMVLHMLKJOLKSQMXRTKJDMPGNPP
UKDGHHXBNLQJMDELOPSGRTLNAQIFF
UJKDMFEOIPSKLFHLDDWNDSLOUFCSL
LPEJKDNFKLRUHFLLSMJCGBQKTDLRI
LKDHGRYUIEPKDSIWLGORRMFILEKMD
```

**Figure 5.2** Sample for Groups 1A or 1B (see below for more details).

A. Cross out the Es in the passage shown in Fig. 5.2

B. Cross out adjacent letters that are the same

The letters can be chosen so that unwanted ones differ from the target by larger or smaller amounts. Thus Es are difficult to distinguish from Fs; by contrast, Os are easier to distinguish from Xs, for example.

## Group 2: tests which stress the motor output

Clearly, some sensory input is necessary, but this, and the amount of thought required, are small in comparison with the motor response required, which is often some form of manipulative skill.

A. Put a dot into the centre of each circle (Fig. 5.3)

The test would consist of about 20 'lines', each of about 20 linked circles.

B. Threading beads onto a string

In case A, the task is easier if the targets are larger, are closer to each other, and more regularly arranged. In case B, the size of the beads and the hole through them can be varied.

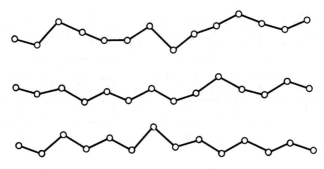

**Figure 5.3** Sample for Group 2A (see p. 70 for more details).

## Group 3: tests stressing coordination between sensory and motor components

*A. Copy each symbol accurately (Fig. 5.4)*

Different degrees of difficulty could be introduced here — both in the shape of the symbol and the standard of accuracy that would be required for a correct answer.

*B. Substitute one symbol for another*

Letters A–F could be paired with six of the symbols in Fig. 5.4, and then a string of random letters (BCEPADDB, for example) would be required to be encoded by use of the relevant symbol. The level of difficulty could be varied in the same ways as with test (A) of this group.

## Group 4: tests stressing short-term memory

The general format of one version of the test is the same as that for tests (A) and (B) in Group 1. The difference is that each line has to be searched to establish the presence or absence of a target set of letters (the order of the target set is not important, only whether or not the whole set is present). Whilst scanning the letters, this set has to be remembered, and it is changed for each test. The target length is generally 2, 4, or 6 letters, requiring low, moderate, and high amounts of short-term memory, respectively. For example, the target letters might be 'MW', 'OWPS', or 'QXTLDG'.

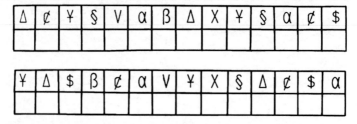

**Figure 5.4** Sample for Group 3A (see p. 71 for more details).

A simpler version of this kind of test would be for the subject to be given a sequence of random digits, and to have to repeat the sequence, either in the order given or in the reverse order.

## Group 5: tests stressing reasoning or some other form of thought

### A. *Logical reasoning (1)*

The sequence 'AB' or 'BA' is given randomly. A statement about this sequence is given, for example: B does not follow A; A precedes B. The subject is required to answer if the statement is true or false.

### B. *Logical reasoning (2)*

A set of statements is given about the relationship between A, B, C, D, E , for example: A is greater than B; B equals C; C is greater than D; D is less than E. The subject is required to state the relationship between A and E, choosing from 'greater than', 'equal to', 'less than', or 'cannot say'.

### C. *Syllogisms*

Accepting the truth of two premises, one has to choose the most appropriate conclusion from a set of four possibilities. For example:

- Premise one: no foods are poisons
- Premise two: some drugs are poisons

Therefore:

1. No foods are drugs
2. Some drugs are not foods

3. Some poisons are not drugs

4. No drugs are foods

D. *Mental arithmetic*

Take the example: $47 + 93 - 28 = ?$ This is as obvious as it is tedious! Clearly, varying levels of difficulty could be introduced into the test. The final two groups of tests cannot be illustrated.

## Group 6: measuring speeds of response (reaction time)

A. *Simple reaction time*

When a light is switched on, or a note is sounded, a button must be pressed as quickly as possible; the delay in doing this is measured. In this case, an 'error' is not possible, but delays over about 1 second could be counted as such. In real life this might mimic having to apply the foot brake for an emergency stop in a car.

B. *Choice reaction time*

A modification of test (A) is for one of a set of buttons to be chosen at random and illuminated. The illuminated button must be pressed as quickly as possible to extinguish the light. 'Error' now can refer to an excessive delay in pushing the right button or to pushing the wrong button.

## Group 7: vigilance tests

Due to the very nature of vigilance, the tests to measure it must take much longer than is required for the other tests. A test session lasting 30 minutes or more is not unusual (but even this need not replicate a real situation, where the time span might be much longer). The test requires the subject to indicate when a signal is abnormal or missed. Thus a stream of pulses lasting 1 second each and given at 10-second intervals could be the 'background'. They could be sound pulses or pulses on a screen, for example. The 'signal' being sought could be a missed pulse, one that was shorter or longer than the standard value, or one that appeared too soon or too late. In addition, the size and frequency of the signals could be varied. The percentage of signals that is detected correctly is generally measured.

## 5.6 The usefulness of such tests — and still more problems

A great advantage of these tests — one that was so difficult to realize with 'real' tasks — is that sets of the tasks of equal difficulty can be devised. Therefore, if the external conditions (temperature, lighting, and comfort of the volunteer, for example) are standardized, the results can be assumed to reflect the mental performance of the individual.

Even so, the tests remain susceptible to factors that can distort the results. Each session normally consists of several types of test, together with a vigilance session of, say, 30 minutes. The total session might take up to one hour. This not only makes it unlikely that employers will allow many such sessions (and this is a severe disadvantage if circadian rhythms are being investigated) but it also means that the tests, particularly by virtue of being repetitive (and necessarily so), are boring for the subject. Also, performance testing is highly susceptible to the effects of the subject becoming distracted; even though the tests and conditions of testing can be standardized, the motivation of those taking them cannot. As an example of this, we found that subjects liked to accompany each test session with music, but listening to slow or fast music altered how they performed the tests. Having discovered this, we forbade music during test sessions!

Another problem is that the tests show practice effects; that is, for the first day or so of testing, performances tend to improve (so masking any underlying daily rhythm) as the subjects are getting used to the tests and are developing their personal way of tackling them. To counteract this problem, subjects are given sufficient practice tests before the experiment to ensure that this effect has worn off as much as possible.

All these difficulties must be taken into account when the results of tests of mental performance are interpreted.

## 5.7 What do these tests tell us about circadian rhythms of mental performance?

In spite of all these difficulties, results have shown that mental performance displays circadian rhythmicity. Some typical examples are shown in Fig. 5.5. For obvious reasons, most tests have been performed only during the daytime.

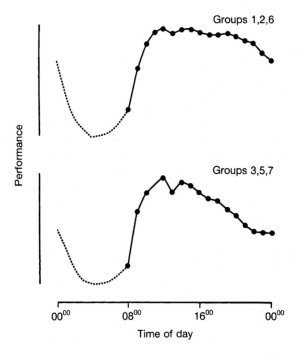

**Figure 5.5** Relationship between time of day (when awake) and performance at different group tests. Improved performances shown upwards. (For definitions of groups, see pp. 69–73.) Dashed line indicates the kind of results to be expected if measurements were made during the night.

With the exception of tests in Group 4, performance rises after waking and falls during the evening. Between these times it shows either a slight rise or a general plateau (tests of Groups 1, 2, and 6) or a slight decline after a peak at about noon (Groups 3, 5, and 7). In most subjects, there is also a temporary decline in the early hours of the afternoon — the 'post-lunch dip'. When subjects are tested at night their performance is generally rather poor. In short, the results of most tests are very similar to 'real tasks' measured 'on site' (but measured with much more difficulty and inconvenience, see Fig. 5.1). This similarity of daily profiles with real tasks is also good evidence that the performance tests are giving meaningful results.

It will be noted that, in general, the performance rhythms of Groups 1, 2, and 6 are similar to that of body temperature, and this fits in with the view that such performance reflects the activity of the brain and the nervous system — an activity that is influenced by body temperature and the speed of metabolic processes in general. One marked exception, however, to this similarity of profiles is provided by those tasks that have a large short-term memory component (Group 4 tests). These tasks appear to be done better at night, with a general decline throughout the daytime to lowest values in the early evening. The explanation for this very different timing of the circadian rhythm of short-term memory is uncertain. It has been suggested that an increasing body temperature and brain activity reduce the brain's ability to store a piece of information without modifying it in any way.

The parallelism between mental performance and body temperature is not exact; mental performance in the evening declines more rapidly than does body temperature. This difference is more marked with rhythms of Groups 3, 5, and 7 (see Fig. 5.5). It can be described as due to increasing 'fatigue'. In general, fatigue increases (and mental performance worsens) as the amount of time spent on a task increases, particularly if we have lost sleep or the task is repetitive and boring (see also Fig. 1.3). Vigilance (Group 7) is therefore particularly susceptible to fatigue. As a countermeasure, it is found that an individual generally benefits from continual changes in the type of task being performed, and from continual updates on how well or badly a task is being performed. There is also benefit to be gained from a break or, better still, a short sleep.

This consideration of fatigue raises the idea of 'double negatives', or even 'triple negatives'. It has been encountered before when 'alertness' was considered in Chapter 3, and these concepts were incorporated into the two-component model of alertness (see Fig. 3.1) in which there was a circadian component and a component that declined with time awake. Considering only that component of mental performance which parallels the rhythm of body temperature, it would be predicted that there would be a trough between 04:00 and 06:00. However, if a subject worked at night and had also been awake a long time, then performance would be expected to decline more due to the other component, time awake. This decline would represent a 'double negative', and was built into the model shown in Fig. 3.1. If, in addition, the subject had lost sleep (and this is often the case during spells of night work), then a 'triple negative' circumstance, with the potential for marked deteriorations in performance, would exist. Such circumstances should, of course, be avoided wherever possible.

In an emergency, the effects of daily rhythms and fatigue are overridden to some extent by the body's emergency response system — if you smell burning in the middle of the night, you will respond without consulting any body clock! Even so, there can be no guarantee that responses will be as appropriate and as well carried out when they are performed in the middle of the night rather than in the middle of the day.

Normally, we are not stressed and so, particularly in repetitive and boring situations, the effect of our body clock and fatigue upon mental performance become relevant. However, any effort to overcome these troughs of performance can, in the long term, become a source of stress (see also Chapter 12).

## 5.8 Making the tests relevant to real tasks

Even though the rhythms in performance measured by the tests and 'real' tasks are similar (compare Figs. 5.1 and 5.5), it seems possible to argue that the tests described so far — for example, crossing out duplicate letters from blocks of random letters, answering questions of logic relating to the sequence of two letters, and copying symbols — are, nevertheless, no more than tests produced by psychologists for use by psychologists, and that their relevance in the field remains to be proved. In attempts to deal with this type of criticism, the basic tests have been 'upgraded' in several ways. Examples are:

- Several types of test (and therefore several components of mental performance) have been combined. For example, in one test, subjects are required to insert ball-bearings into holes in a cylinder. The holes are of slightly differing sizes, the ball-bearings also are slightly different in size, and the cylinder is rotated irregularly. Such a test combines elements of Groups 1–3. As another example, subjects have to draw an irregular shape — the shape being seen as a reflection in a mirror. This combines elements of Groups 1–3 and 5 with a fair degree of exasperation!

- Several tasks are required to be performed simultaneously, and the subject is required to distribute time between them as efficiently as possible. This is supposed to mimic the stress of the decision-making process. At times, it seems to those taking part to resemble a bad dream, as they are required to be vigilant in spite of distractions caused by other unfinished tasks.

• Complex, computer-based games have been used, so that scores for manipulative skill, decision taking, vigilance, reaction time, and so on can all be calculated as the game proceeds. Clearly, games can be devised that test more subtle strategies and complex manoeuvres. At a simple level, they might place emphasis on speed and/or accuracy, and explore how an individual devises a 'trade-off' between these two opposing aims. Taking this idea to an extreme brings us to 'virtual reality' driving tasks and flight simulators, both of which aim to duplicate real-life situations. If used repetitively at different times of the day, they could assess changes due to daily rhythms.

By now, however, we have almost come full circle in attempting to measure mental performance. We originally considered the measurement of real tasks 'on site', but this was prone to disturbances caused by external factors. We then considered simple tests to overcome some of the difficulties, but these were often not very 'relevant'. So we made these tests more complex, in order to increase their relevance, but in so doing we produced ones which were so sophisticated as not to be widely available, due to cost and personnel requirements. Somewhere in this circle, it is hoped, there is some test or combination of tests that is not only relevant to the real task under consideration but also can be standardized more easily.

## 5.9 Measuring mental performance indirectly

Is there an alternative to this, at times, unhappy compromise between 'relevance' and 'simplicity', and the associated problems?

Some scientists have suggested that there are daily rhythms which mirror those of mental performance, but which are not as difficult to measure or as prone to interference — body temperature and adrenaline excretion are often cited in this context. The parallelism between these two rhythms and performance has been illustrated earlier (see Fig. 4.1, for example), but it must always be remembered that this parallelism (or oppositely timed rhythms in the case of short-term memory) has not been tested under all conditions. Therefore, at times, it might be wrong. Nevertheless, it is the case that, during shift work and after time-zone transitions, rhythms in body temperature and adrenaline have been used in addition to, or even in place of, measurements of performance at 'real tasks' or at the simpler tests devised by psychologists.

Other rhythms — those of heart rate (see Fig. 4.4) and oral temperature — have been used also. As already described ('Now try this' section at the end of Chapter 3), oral temperature gives faulty results if the subject has recently been talking or drinking, and it cannot be measured when the subject is asleep. Heart rate is unsatisfactory because its external cause is so much stronger than its internal cause — so its rhythm tends to reflect the individual's lifestyle and environment much more than that of the body clock.

# Now try this: measure your own performance

You might like to try and measure your own performance. This requires care and patience in the preparation, performance, and marking of the tests, but can be most rewarding. The comments below are intended as guidelines only. More details can be found in the chapter.

## 1. Choice of tests

With sufficient ingenuity, all the tests except those in groups 2B, 3A, and 5C could be produced by a computer. Many tests are best printed out on paper, but those in Groups 5–7 can be run on the computer screen equally well. An alternative would be to measure your score in an arcade-type game on your computer.

## 2. How many tests are required?

At least 20 sessions, spread over several days, are required to remove the major effects of practice. The investigation itself requires the test sessions to be repeated as often as possible to reduce the effects of variability. This might, for example, be on 5 occasions at each test time, or 5 subjects studied at each test time. So, if you plan to measure performance at 8 times of day, then 40 complete test sessions, in addition to the practice sessions, will be required.

## 3. Performing the tests

Remember that performance is susceptible to the effects of external conditions, the mood of the subject(s), and sleep loss, as well as to rhythmic changes. So it is most important to standardize the frame of mind of all concerned, as well as the conditions for doing the tests.

## 4. Marking the tests

Generally, the speed of performance (that is, the number of questions done in the time allotted or the time taken to complete the test) is scored rather than its accuracy. Note that a good performance means that more sums have been done in a set time or that less time has been taken to complete the test.

Individuals can differ widely in their scores and this makes direct comparisons between them difficult. (It can also upset some subjects to find out that they have performed below average for the group.) One way to deal with this is to express each result from an individual as a percentage of the mean value over all tests for that person; in this way, different subjects, whatever their relative abilities, will all vary about the same mean value of 100 per cent. Also, if you are looking for rhythmic changes in a single volunteer, then it is the change in performance during the course of the day, not the absolute level of performance, that is most important.

Beware also of another problem. If you are scoring the tests yourself, rather than getting the computer to do it for you, then always try to do so at the same time of day, otherwise you will be introducing into the results another source of variability — caused by time-of-day effects on your scoring ability!

Good luck!

# Chapter 6

## Food, metabolism, and the removal of waste

The circadian rhythm of metabolism is part of another group of rhythms, which includes also rhythms in the intake and digestion of food, and in the removal of waste material from the body. Metabolism, the processes by which food is broken down chemically to release its energy, has already been mentioned in Chapter 4. There, it was pointed out that metabolism is essential for enabling physical exercise to take place, and that it requires the uptake of oxygen into the body.

## 6.1 Is it time to eat yet?

The animal and plant kingdoms provide examples of a wide range of eating habits. For example, fungi 'eat' continuously, and cause decay at all times. Many parasites probably do the same, feeding on their host throughout the 24 hours. Animals such as sheep and cattle tend to forage throughout the daylight hours, whereas predators hunt whenever their prey is around, day or night. Green plants photosynthesize (make food by using the energy of sunlight) whenever it is light. Mosquitoes and other blood-sucking insects feed whenever they are active, often in the evening.

Where in this wide range of feeding patterns do we humans fit? We appear to eat periodically while we are awake, the times of eating being dominated by social factors and the structure of our day. For example, in many countries, there is legislation to ensure that coffee breaks, tea breaks, and a lunch hour are all incorporated into work hours. After work, we tend to have another large meal, and then a snack before retiring. We do not need to eat that frequently, and the high incidence of obesity in many Western populations indicates that the number of calories that we take in is excessive compared to our general activity levels.

Part of the reason for our eating habits is not only that they fit in with work schedules but also that they serve a social function. Unlike babies, we do not eat or drink whenever, and just because, we feel hungry or thirsty. Meals can serve as a focus for certain social activities. Consider, for example, the 'Sunday roast', the wining and dining of lovers, and even the 'working breakfast'. Fluid intake too, in the form of wine, beer, spirits, and cups of tea or coffee, serves this social role as much as a biological one.

In addition, there are considerable differences between individuals. Some people miss breakfast, others have their main meal at midday rather than in the early evening — differences determined by personal preferences (or schedules) as well as cultures.

If the timing of our intake of food and fluid tends to be determined by social factors, what about the type of food and fluid that we take in on a particular occasion? Again, social factors are important; morning coffee, afternoon tea, and dinner parties describe not only the timing of the occasion but often also the amount and type of food and drink that we will be given. There are other factors too, of course, including food availability and the amount of time available. However, fridges, tinned food, microwave ovens, and 'instant meals' all reduce the importance of these factors in most circumstances.

Even so, biological factors also influence our food and fluid intake. We drink if we feel thirsty, and thirst is brought on by a fall in content of body water, indicated by a rise in the concentration of body fluids. A good example is the increased drinking that occurs on a hot day as a result of sweating. The sensation of thirst originates in a region of the brain called the hypothalamus, and this regulates fluid intake during the course of each day.

The hypothalamus is also involved in making us feel hungry, probably related in some way to the falling level of glucose in the blood. But the amount of food that we take in is not determined much by the amount of time that has elapsed, or the amount of energy expended, since the last meal; rather, it is determined by social factors. We do regulate food intake, of course, but this is in the longer term, over days and months. For many people, food intake during the week is slightly less than that required to maintain body weight, but this is made up during the weekends by larger meals, often accompanied by alcohol, as part of social occasions. Those who are in physical training and expend more energy will start to eat more as the days go by; and, in contrast, people who have put on weight will begin diets and other slimming regimens.

Is there any evidence of a daily rhythm in the desire to eat? This is a very difficult issue to resolve. In many experiments upon individuals isolated from environmental time cues (free-running experiments), food intake — breakfast, 'elevenses', lunch, and so on — continues to be distributed normally during the course of the waking span. In these cases, food intake might be linked to the body clock and to a circadian rhythm of energy expenditure, but it might also be 'habit'. Even in those cases where the sleep–wake cycle becomes particularly long and disassociated from the rhythm of body temperature (see Chapter 2), a conventional distribution of meals **within the waking span** continues. This result also indicates that habits rather than the body clock might determine when during the 24 hours we eat. Indeed, many subjects indicate clearly how important habit is, stating that they eat because they believe it is the 'right' time to eat. In summary, there is very little evidence that the body clock exerts a strong effect upon food intake.

This idea that food intake is due to habits, in turn due to conforming to social norms, can become a problem for night workers. Their eating patterns will be considered in more detail in Chapters 12 and 13. Here we state that their eating patterns are abnormal, with more snacking of 'convenience' or 'junk' food during the 'lunch' break in the middle of the night. On other occasions, when they feel tired after work, they know they should eat (a rumbling stomach and hunger pangs tell them this) and yet they have no appetite. Jaded palates need some titillation, and yet another 'snack' might not be the answer.

These problems cause concern, but dealing with them effectively is difficult if the cause of the abnormality is not fully understood. Thus, the advice to the night worker, and to the employer, would be very different according to whether the abnormal pattern of food intake were food availability, food palatability, biological desire or habit, for example.

## 6.2 Indigestion and constipation

After eating the meal, the food must be digested. This requires the coordinated release into our gut of digestive enzymes and fluids that are then mixed with the food. These enzymes break the food down into its constituent sugars, fatty acids, and amino acids. While being digested, the food is passed slowly through the gut by the action of smooth muscle in the gut wall, and foodstuffs, water, and salts are absorbed into the bloodstream. What remains, after conservation of most of its water content, is converted into solid faeces for elimination.

There is very little evidence to indicate that these processes cannot be achieved by the body equally well at any time of day or night. Our personal experiences — for example, of indigestion following a large meal late at night — are difficult to interpret. Does the indigestion result from the timing of the meal (the point in question), its size or richness, the wine that accompanied it, or the act of staying up late? In the same way, when individuals fast during much of the day for religious reasons, the argument that it is difficult to concentrate in such circumstances must be weighed against the fact that it appears that mechanisms of metabolic regulation enable blood glucose to be controlled for quite extended periods.

There is evidence that the concentration of acid in the stomach is highest in the night, and this explains why pain associated with gastric ulcers tends to be worse then. However, this need not mean that there is a circadian rhythm of acid secretion by the stomach — the stomach is empty at this time, so the acid that is released by the stomach remains free in it rather than mixing with food.

As the residue that is left after foodstuffs have been absorbed into the blood passes towards the anus, the desire to defecate results. In herbivores particularly, but also in many animal and human babies, defecation takes place more than once per day. In health, constipation should be rare. Its high incidence nowadays in many Western cultures is a reflection of the fact that many people have too little roughage in their diet. In adult humans, the decision whether or not actually to defecate is under voluntary control. Most of us defecate once a day, and at a convenient time. There are reflexes in the gut that operate to stimulate defecation after food has entered the stomach. This reflex is likely to be most marked after a period of fasting, and this would appear to account for the morning visit to the toilet after breakfast in many people. However, this routine can become disorganized after time-zone transitions or during night work, and this disturbance can become inconvenient and even socially embarrassing, particularly if we have a busy and highly ordered lifestyle. Such a lack of an immediate adjustment to altered sleep–wake schedules suggests that the body clock plays some role in this process, though what form this link might take has not been studied.

## 6.3 Making use of our food

The biological aim of the meal is to enable food to be taken into the body for its growth and repair. After digestion, the sugars, fatty acids, and amino acids are absorbed into the bloodstream. The amino acids are used to build and repair body tissues, whilst the fatty acids are stored as fat in adipose tissue, and the sugars as glycogen in the liver and muscles.

Several hormones, particularly insulin (the hormone that is deficient in sugar diabetes), control these processes, and glucose is a key substance in the regulation of metabolism. Insulin prevents a rise in blood sugar, which would adversely affect brain function. It does so, not only by promoting the laying down of the storage form of glucose, glycogen, in the liver and muscles, but also by promoting the metabolism of glucose for energy. Insulin also promotes the deposition of fat in the adipose tissue.

Insulin is released into the bloodstream from the pancreas, and the amount of insulin released in response to the rising sugar level in the blood, together with the sensitivity of the tissues in responding to this insulin, depends upon the time of day. The system works more effectively in the daytime than at night, and works best of all in the morning. That is, the body is better able to deal with glucose in the daytime than at night. Energy is required at night, however, for growth and repair, and also to keep the body 'ticking over', even though we are asleep. Energy comes now from the breakdown of fats. This change in emphasis of metabolism, from glucose towards fat, is brought about by the different hormone concentrations that are found in the blood during the night (see Fig. 6.1).

Important in this metabolism of fat to provide energy is the decreased secretion and effectiveness of insulin, but important also are the raised secretions of three other hormones — growth hormone, glucagon, and cortisol. At least part of the increased secretion of these hormones is under the influence of the body clock.

We can therefore say that the body clock exerts an influence over the metabolism of food. This is believed to explain the observation that, in one experiment, there was a difference between the effects of eating a single meal in the morning or evening. (The meals were of equal size and calorie content at whichever time they were eaten.) Subjects tended to lose weight when they had eaten the meal at breakfast time, but not when they had done so at dinner time. However, before assuming this is a successful method for dieting, it must be stressed that the effects were small, and that the difference might have reflected a greater energy expenditure in subjects who had eaten breakfast (having been 'set up' for the day).

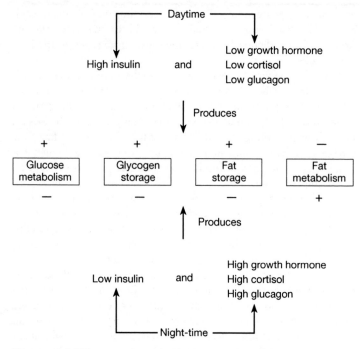

**Figure 6.1** Differences between the hormonal regulation of metabolism in the day and in the night. +, stimulates; −, inhibits.

Nevertheless, this finding does indicate that food metabolism is under the control of the body clock to some extent and, once again, turns attention towards those who work at night. It might be that not only does their desire for food change (see above) but their ability to metabolize it also varies. This will have further implications when advice on diet is given to night workers.

## 6.4 Our rhythmic kidneys

Linked to the intake and metabolism of food and drink is the elimination in the urine of excess water and salts and of soluble waste materials.

All substances that appear in the urine show daily rhythms in their rates of excretion and these are due to the effects of habits as well as the body clock. The internal cause of the rhythm can be established in the normal manner — by making the intake of food and water, the individual's posture, and the environment constant throughout the 24 hours (see Fig. 1.4). The rhythm that persists in these circumstances indicates the effect of the body clock, and the difference from the rhythm when living normally is due to external causes. Such experiments indicate that the relative importance of internal and external causes depends upon the substance. For example, the internal cause of the rhythm in potassium excretion is quite large; for calcium and water excretion, the external cause is dominant (see Fig. 1.4); and, for phosphate excretion, the two causes are of similar strength.

The external cause of the rhythms of urine formation is twofold — our diet and changes in posture. The effects of meals are easy to understand; we drink and eat in the daytime, but not at night, and so there is an excess of fluid and salts in the body in the daytime, and a deficit during the night. With regard to the effects of posture, lying down increases the return of blood to the heart and brain from the legs. (It is for this reason that lying down when you feel faint is a good idea.) This increased return of blood is monitored by receptors that signal to the brain the amount by which the heart is filling. These receptors interpret the increased return of blood when lying down as a sign of an increase in blood volume. This causes changes in the release of two hormones, antidiuretic hormone and atrio-natriuretic hormone, as a result of which the kidneys excrete more water and sodium. Such a reflex effect occurs also in outer space where the zero-gravity environment means that blood no longer pools in the limbs. The resulting increase in urine flow severely tested the durability of the spacesuits worn by the early astronauts!

The aforementioned reflexes guard against an excessive blood volume, associated with which would be an undesirable rise in blood pressure. However, they, and kidney function in general, are damped down at night. This modification of function at night is due to several factors, amongst which are altered activity in the nerves going to the kidneys and a reduction in the blood supply to the kidneys. It is a very good example of the internal cause of renal rhythms (the body clock) overriding the external cause (a change in posture). The result is important since it slows down the rate by which the bladder fills during the night, so allowing us to sleep longer. Without this change, the pleasures of a 'night cap' just before bedtime would be considerably reduced!

In the daytime, by contrast, despite an upright posture, urine flow increases due to the effects of the body clock (acting via the nerves to the kidney and the increase in blood supply to the kidneys) and to the intake of fluid. Daytime sleep, when the body clock does not damp down these postural reflexes, is often curtailed by the rapid filling of the bladder that this postural reflex produces — as can be vouched for by night workers who need to sleep at this time.

A reduction in the strength of the postural reflexes at night is not observed in some renal and heart disorders; then, the rate of urine flow on lying down at night is as great as when doing so during the daytime. This is diagnostically useful to the physician. In babies and some old people also, the circadian effect of the body clock on renal rhythms is weaker and a reduction in urine flow at night is not seen. Particularly for old people, this, coupled with a tendency to have a 'weak bladder', can have distressing consequences.

The strong internal cause of the rhythm of potassium excretion in the urine is because it is partially controlled by the hormone cortisol.

A knowledge of the rhythms in urine formation can be important when treatment with drugs is concerned, since some are removed from the body in the urine. The degree of acidity of the urine shows a daily variation, nocturnal urine being most acidic. Many drugs are acidic or alkaline, and this characteristic affects their removal in the urine; acidic drugs are excreted more easily in the daytime and alkaline ones, at night. Since drugs are costly, it might be desirable to give them when they are less readily removed by the kidneys. This timing will reduce the amount of drug required, and so, the cost of treatment.

Giving less drug might have another advantage, since some drugs, whilst having other beneficial effects, are toxic to the kidney. Thus, there is always the risk that the kidneys will be damaged while excreting the drug. Since the kidneys produce the largest volume of urine in the middle of the day, any drug that is being removed will be least concentrated in the urine, and so be least likely to cause damage, at this time.

The physician or doctor will have to bear both these factors in mind when deciding if the time when a drug is to be taken is important (see also Chapter 14).

## Now try this: have you eaten yet?

As discussed above, whether or not individuals eat at a particular time might reflect their physiology (appetite or body clock), their habits, and whether they or somebody else will have to do the cooking, as well as social influences and time pressures. In addition, how relevant are these factors when it comes to the type of meal that is eaten? Does the type of meal depend also on the cooking and eating facilities that are available? And do the results depend upon the time of day, whether it is a weekday or a weekend, whether it is a day of rest or work, or whether the work is done in the daytime or at night?

Table 6.1 reproduces one page from a questionnaire that has been designed to assess the factors that determine why individuals do or do not eat. The questionnaire is given every 3 hours during a 'typical week' and consists of 56 pages like the one shown (8 per day × 7 days). A 'typical week' depends upon the detailed nature of the investigation being undertaken.

Discovering what factors influence our eating enables advice that has a rational basis to be given, both to the individuals themselves and to those concerned with the provision of suitable food and eating facilities.

The respondent should not be aware of your detailed reasons for wanting the questionnaire to be answered. For example, if it is perceived that you wish to investigate whether night workers need better facilities, then the answers are likely to be different from those when the workers think you wish to show that the type of meal depends upon social influences or habits. Such a bias can be reduced by explaining your aims as little as possible, even by giving instructions in the form of a carefully worded letter rather than by personal contact. However, without personal contact you cannot clarify any difficulties that a respondent might have — and you are less likely to get any reply at all!

The criticism that different individuals will interpret the same question differently (What is meant by a 'small' rather than a 'large' hot meal? What does a score of '7' mean in questions 9–11?) need not be as great a problem as it seems at first. If the same subject answers the questions on different occasions, then the differences between the responses will reflect the difference between the two situations — and this was the prime aim of the questionnaire. Of course, it is valid to consider if individuals in the same circumstances answer the same questions differently, but this is a different problem and would require a different approach.

**Table 6.1** A questionnaire about eating habits

**TIME: 09:00–12:00**

**Q1** Did you sleep throughout this period?    NO / YES
           (a)    If NO, go to Q2
           (b)    If YES, end of the questionnaire

**Q2** Did you have a meal during this period?    NO / YES
           (a)    If NO, go to Q3
           (b)    If YES, go to Q4

**Q3** Why did you not eat a meal during this period? (Tick as many as apply)
           (a)    I never do
           (b)    I did not feel hungry
           (c)    There was nothing available to eat
           (d)    I did not have enough time to eat anything
           End of the questionnaire

**Q4** How many people were in your group eating together at this meal/snack?
_____ (1 = I was alone; 2 = myself and 1 other; and so on)

**Q5** Why did you eat at this time? (Tick as many answers as apply)
           (a)    I always do
           (b)    I felt hungry
           (c)    To meet partner/friends
           (d)    My schedule dictated I had to eat now

**Q6** Who prepared the meal? (Tick the most appropriate answer)
           (a)    I did myself
           (b)    Partner/friend
           (c)    I bought it
           (d)    Provided by work

**Q7** What determined the kind of food eaten? (Tick as many answers as apply)
           (a)    My appetite
           (b)    My habits
           (c)    Time available
           (d)    Cost
           (e)    Food available

**Q8** Which of the following did your meal/snack contain? (Tick as many answers as apply)
           (a)    Beverage
           (b)    Snack food
           (c)    Cold food
           (d)    Small hot meal
           (e)    Large hot meal

*(Continued)*

**Table 6.1** (Continued) A questionnaire about eating habits

Q9 How hungry were you before the meal/snack? (Ring the most appropriate answer)

| 0 | 1 | 2 | 3 | 4 | 5 | 6 | 7 | 8 | 9 | 10 |
|---|---|---|---|---|---|---|---|---|---|----|
| Not at all | | | | | Moderately | | | | | Very |

Q10 How much did you enjoy your food? (Ring the most appropriate answer)

| 0 | 1 | 2 | 3 | 4 | 5 | 6 | 7 | 8 | 9 | 10 |
|---|---|---|---|---|---|---|---|---|---|----|
| Not at all | | | | | Moderately | | | | | Very |

Q11 After the meal, how did you feel? (Ring the most appropriate answer)

| 0 | 1 | 2 | 3 | 4 | 5 | 6 | 7 | 8 | 9 | 10 |
|---|---|---|---|---|---|---|---|---|---|----|
| Still hungry | | | | | Comfortably satisfied | | | | | Over-full |

# Chapter 7

## Your body clock at different stages of your life

So far, we have considered daily rhythms in healthy adults. Here we will consider the differences from adults that are found in the rhythms of healthy babies and aged people. These are important stages of our lives — too often biology textbooks neglect them and assume we are forever in our youth and prime! We can gain clues about the nature of the clock by a study of these stages, as well as use our understanding of it to offer advice.

### 7.1 The development of rhythms in young children

The newborn child does not show marked 24-hour rhythms, as any parent will affirm! The baby's feeding habits, which directly concern a nursing mother, and its waking and crying habits, which can concern the whole household, tend to be distributed much more evenly throughout the 24 hours than is found in the adult (see the sleep profile of a newborn child in Fig. 2.4). As the baby grows up, the strength (amplitude) of the 24-hour rhythms begins to increase, with more sleep being taken at night and less in the daytime, and with feeding becoming more frequent in the daytime than at night. However, fully established daily rhythms of normal amplitude — that is, ones with markedly uneven distributions of sleep and feeding between the night and the day as are found in adults — do not develop until about five years of age.

Not all rhythms develop at the same rate and that of the hormone, cortisol, appears to be one of the last to develop. Due to their largely external cause, many rhythms appear to develop at the same rate as the sleep–wake cycle, urine production and heart rate being examples of this.

It used to be thought that the neonate showed no daily rhythm of body temperature. Recent work, though, has shown that healthy, full-term babies living in a normal environment with an alternation of light and dark do possess a circadian rhythm even as early as the second day after birth (Fig. 7.1, left). However, it has a low amplitude and it is not timed to coincide with the light–dark cycle as accurately as in adults. That is to say, the times of peak vary much more between babies compared with adults. By the fourth week of life (see Fig. 7.1, right), the amplitude of the temperature rhythm has increased, although there are still no clear circadian rhythms of activity, heart rate, or blood pressure.

Moreover, in premature babies living in the constant environment provided by an incubator (because of respiratory problems and/or jaundice), a free-running circadian rhythm of body temperature is again present. Though it is weak, its presence is clear evidence of the existence of a functional body clock.

In rodents also, the newborn pups show weak temperature rhythms, but it has been possible to study the electrical activity of the cells that make up their body clock (see Chapter 8). What these studies have shown is that the individual cells produce a circadian rhythm — as they do in the adult rodents — but that the cells tend to act independently of each other and to be less influenced by the environment. If this applies also to human babies, then it explains their poorly developed 24-hour rhythms; the outputs from the cells are not fully synchronized with one another, so the individual rhythms tend to cancel each other out. The early stages of life in humans are associated with a massive increase in the number of connections between brain cells. If the cells of the body clock show this also, then, as the interconnections form, so too does the interaction between cells, with their output increasing in synchrony. This would account for the growth in strength of body rhythms in the time immediately after birth.

As the strength of the body clock increases, not only body temperature but also other aspects of the infant's physiology and biochemistry will show an increase in the strength of the daily rhythms. These include the rhythm that is of most concern to others — the sleep–wake cycle!

Presumably, the ability to respond to the environment and adjust the timing of the body clock (for the light–dark cycle to act as a zeitgeber) develops at about the same time, though there is some evidence that it is an independent process. Thus, in a few cases, it has been found that, in children studied in a normal, rhythmic environment,

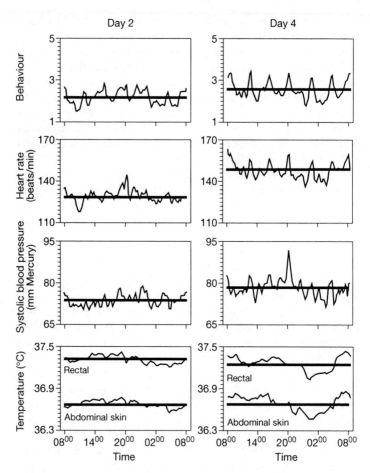

**Figure 7.1** The daily changes in activity, heart rate, blood pressure, and temperature measured rectally and on the skin in a group of healthy full-term babies studied 2 days (left) and 4 weeks (right) after birth. (Behaviour score — 1 = deeply asleep; 2 = lightly asleep; 3 = awake, drowsy; 4 = awake, alert; 5 = awake, crying.) Horizontal lines indicate 24-h means.

the sleep–wake cycle shows an intermediate stage between a lack of circadian rhythmicity and a regular profile with a timing that is adjusted appropriately to the light–dark cycle (Fig. 7.2). This intermediate stage appears to comprise a free-running rhythm with a period that

**Figure 7.2** Distribution of sleep (black bars) and wakefulness in one infant. Note that daily rhythms are absent or weak for the first 8 weeks and then show a period greater than 24 hours between 9 and 17 weeks.

differs from 24 hours; in these cases, the rhythmic output of the body clock appears to be temporarily ahead of the ability to adjust its phase.

## 7.2 The role of the environment

The development of the body clock normally takes place in a rhythmic environment, and it is almost certain that this plays a role in the overall process. Suppose that the baby wakes at random times — in order to feed, for example — there will be some times during the 24 hours when its environment is less stimulating than others. This rhythmic environment might be by parental design or due to natural events. Parents might decide to put the baby 'into a routine', and this would mean imposing some sort of 24-hour cycle upon the child.

This routine might include the amount of light, noise, and general stimulation the child received during the daytime and the lack of these at night. It might also involve feeding times; during the daytime, meals would be given more readily in response to the child's cries than during the night. Even if the child were fed on demand, its environment and parents' sleep would reduce the likelihood of the baby waking and being fed in the night compared with in the daytime.

When the process of promoting the development of circadian rhythmicity is considered, is putting a baby into a routine advantageous to the child (ignoring the likelihood that it is advantageous to the parents)? In one study, body weight increased more, and general development progressed more rapidly, in a group of premature babies living in a hospital nursery where rhythms of light–dark and noise–quiet had been imposed, than in a group who had been in a similar ward but with less emphasis on the differences between night and day. Perhaps, therefore, a rhythmic environment is beneficial. In the babies referred to in Fig. 7.1, there was evidence of a direct effect of the light–dark cycle — sleep in the dark causing greater falls in body temperature, heart rate, and blood pressure; and waking activity in the light raising them more. These results also point to the benefits of imposing a rhythmic environment upon newborn babies when the development of daily rhythms is considered.

## 7.3 The role of lifestyle

It must not be assumed that the finding of weak 24-hour rhythms in newborn babies means that they have little rhythmicity. On the contrary, babies show a number of rhythms — in their sleep–waking activity, their feeding requirements, and their bladder-emptying, for example — but these have a much shorter period length, of 1–4 hours; for this reason, the rhythms are called ultradian. These ultradian rhythms are present in adults too, although their amplitude is less than that of the 24-hour rhythm. The clearest example of an ultradian rhythm in adults is the alternation of the REM/non-REM cycle of sleep (Chapter 3), but ultradian rhythms in mental performance have been described also. In addition, we have already seen that it is common to feel tired about lunchtime (the 'post-lunch dip'); this is about 12 hours after retiring, and so is another example of an ultradian rhythm.

In neonates, the ultradian rhythms seem to derive from the baby's lifestyle — which means that there is no need to search for an internal clock that shows ultradian periods! In the study (above) of healthy neonates living in a normal light–dark environment, the rhythms of blood pressure, heart rate, and behaviour (the sleep–wake cycle) all showed a period of 3 hours (equal to that of the routine feeding and care that the babies were receiving), whereas the body temperature had a rhythm with a period of 24 hours. These ultradian rhythms were present on the second day after birth (Fig. 7.1, left), but were even more marked by the fourth week (Fig. 7.1, right). Indeed, by the fourth week, some of the effects of these 3-hour rhythms were present in the body temperature record also (superimposed upon the 24-hour rhythm).

## 7.4 The interaction between nature and nurture

The ability to develop circadian rhythms is part of our genetic make-up (see also Chapter 8), but it is only in a normal, rhythmic environment that this potential is fully realized. In environments in which there is little daily rhythmicity, daily rhythms do develop, but the amplitudes of the rhythms develop less quickly. A rhythmic input from the environment, one with a period of 24 hours, is important also as a zeitgeber, so that the phase of the circadian rhythms can be adjusted appropriately. Additionally, there are external causes of rhythms — not only the light–dark cycle (a circadian input) but also the effects due to sleeping and being woken for food and routine care (an ultradian input). As with many other aspects of development, there is a mixture of 'nature' and 'nurture'.

Ultradian rhythms last into adulthood but become progressively more overshadowed by the strengthening circadian rhythms. The end result is that components with a 24-hour period are dominant by the time the child has reached his or her fifth birthday. In some children, the development of daily rhythms is poor. An underdeveloped renal rhythm may mean that the production of urine at night is considerably higher than in normal children, and this can lead to the problem of bedwetting (although emotional problems are often a more likely cause of this).

The development of daily rhythmicity is linked to the fact that most of us live according to a 24-hour schedule. However, in some cases, it is not only the 24-hour rhythms that are stressed. A 12-hour day is common in those countries where an afternoon siesta is taken and is also practised by watch-keepers on merchant ships working a shift pattern of 4 hours of work followed by 8 hours of leisure and sleep, the combination being undertaken twice each 24 hours. A sleep profile for someone on such a shift system is shown in Fig. 2.4.

## 7.5 Rhythms in old age

As we grow older, our whole body changes in many ways. We develop wrinkles and our skin loses its firmness; we become less active physically and mentally; and we might become less able to deal with disturbances induced in our bodies by the environment, including sleep loss, jet lag, and the difficulties associated with night work. The use of skin creams and living a physically and mentally active lifestyle will, to varying extents, retard this deterioration, but they will not prevent it.

One of the most marked changes is a decline in secretion of several hormones, particularly growth hormone and the sex hormones (see Fig. 3.4), the latter being due to a declining response of the ovaries or testes to signals coming from the brain. It seems that brain function declines also, and the combination results in an impaired effectiveness of many systems that are responsible for controlling the body. For example, the maintenance of blood pressure when standing up from a lying position is achieved less rapidly as we age; as a result, the blood supply to the brain is decreased for longer and there is a greater tendency to feel dizzy momentarily on standing. Another example is that of temperature regulation. Older people are more susceptible to cold. This is partly because they can be thinner and less active (and so are more likely to become cold), partly because the information about a fall in blood temperature or a cold skin is acted upon less readily by the control systems of the body, and partly because it is more difficult to generate more heat (by muscle activity and shivering).

It is not surprising, therefore, that daily rhythms also change with age. In general, there are three types of change: a decrease in amplitude, a tendency for the rhythms to be timed earlier, and an increase in the day-to-day differences in timing of the rhythms. Why should these changes occur?

The fall in amplitude of some rhythms will result in part from a decrease in their external cause. Taking the rhythms of body temperature, blood pressure, and heart rate as examples, the observed decreases in amplitude would be predicted, since the rises observed in the daytime would be reduced as a consequence of the decreased amount of physical activity. There is the additional possibility that the body clock itself might, like other parts of the brain and body, deteriorate with age. Even though the body clock has been studied comparatively little in older volunteers, there is evidence from constant routine and free-running studies to suggest that the output from the body clock decreases with age, at least in some old people, but less so in others (the 'survivors') (see below). In addition, particularly in extreme old age, there is anatomical evidence that the body clock becomes smaller.

The body clock appears to run slightly faster than in younger controls; that is, older subjects tend to be slightly more of a 'lark' than their younger counterparts. These effects of ageing on the body clock can be seen in individuals who retain an active lifestyle. For example, the times of peak of several circadian rhythms were found to occur about 3 hours earlier (see Fig. 7.3) in an older group of subjects (47–65 years) than in a younger group (20–30 years).

The range of circadian rhythms that shows this change in timing is very extensive, covering rhythms in mental performance and the cardiovascular system, for example, in addition to those shown in Fig. 7.3.

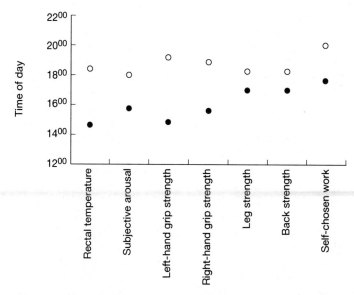

**Figure 7.3** Times of peak of various circadian rhythms in groups of old (solid circles) and young (open circles) subjects.

For instance, veteran cyclists perform closer to their peak performance than do younger cyclists when time trials are conducted in the morning.

Since these changes happen at different rates in different individuals, it can be seen that, as a group, the phasing of rhythms in older subjects might become more variable. Also, the body might respond less well to zeitgebers, and the strength of the zeitgebers themselves might decrease (for example, if the subject is inactive, or lives a rather restricted life socially and physically, or suffers from being only partially sighted). In all these cases, the daily adjustment of the body clock might be compromised.

## 7.6 Insomnia associated with old age

The changes in timing of many of the circadian rhythms are bound up with changes in timing of sleep (see Chapter 3). An earlier waking time need not be a disadvantage, however, particularly if the individual wants to 'get going' early in the morning. By contrast, a decreasing ability to maintain sleep throughout the night is of concern.

Insomnia is a common complaint in the aged, and changed daily rhythms are likely to play a part in this problem. The decline in amplitude of the 24-hour rhythm of body temperature means that the fall of body temperature in the evening is less marked. (This decline might be due to any combination of the following: declining clock activity; declining physical activity in the daytime, so the temperature has not risen to as high a value in the daytime; and declining secretion of melatonin in the evening — the rise in secretion of this hormone in the evening normally accentuating the clock-produced fall in temperature at this time, see Chapter 3.)

There are two additional reasons why sleep might not be as consolidated as it is in younger subjects. First, decreasing mobility and retirement are likely to reduce daytime activities, and there will be a greater opportunity for daytime naps, particularly after lunch. This will reduce the amount of sleep that will be needed at night. Second, sleep at night is more likely to be disturbed by the need to empty a full bladder, since the amplitude of the daily rhythm of urine production also decreases. Therefore, the rate of urine production at night does not decrease as much as it did during youth. The problem might be exacerbated if the subject is taking drugs that combat high blood pressure or fluid retention by promoting the loss of fluid.

All these factors — due to both external and internal causes — will conspire to produce individuals who are less polarized in their lifestyle between daytime activity and sleep at night. Unfortunately, this state of affairs can lead some people to go to bed earlier — to catch up on lost sleep — and this can result in their lying awake in bed even longer!

## 7.7 What can be done?

It is not always such bad news, however, and to understand ways around these problems, we turn to some studies that have looked at 'survivors'. This is the term used to describe those who seem to derive great enjoyment from a productive old age. What can they tell us about ways to overcome some of the problems described above?

Sometimes, it seems that these survivors are just 'lucky', possibly with body clocks that deteriorate less than average. However, in recent studies, older people (aged 50 years and above) who lived fairly active lives were asked to record their habits each day (times of waking, eating, visiting friends, sleeping, and so on) for a 'typical week'. In this way, it was possible to get some idea of the day-by-day variability in their habits. It was found that, as they aged, they continued to be socially active (and even physically so, as far as was possible), but they showed a decreasing variability in their habits. In other words, they became more 'set in their ways'.

There is no definitive explanation of the results. The survivors, blessed with good health and still possessing all their faculties, were young for their years. Perhaps it could be predicted, therefore, that their daily rhythms also would reflect this 'youthfulness', remaining quite stable and high in amplitude. An alternative possibility is that the survivors' regular and active lifestyle exposed them more to zeitgebers, which helped to stabilize daily rhythms.

Even if this latter explanation is not the correct one for these particular subjects, it might be possible to improve the daily rhythms of older subjects — and, therefore, their ability to be active in the daytime and sleep better at night — by making use of zeitgebers.

## Now try this: making your body clock more regular

Particularly as we become older, there is a tendency for rhythms to become less regular. This need not matter, but the following advice is aimed at those for whom it is an inconvenience. This information follows on closely from that given in the 'Now try this' section at the end of Chapter 2.

- **Do** try to get into a routine. Try to strengthen your daily time cues and make them as reproducible from day to day as possible. This means trying to adopt regular times for meals, activities (see below), and going to bed and rising.

- **Do** try to do something interesting each day. Examples of this would include taking a 'constitutional', visiting a friend, or going to a meeting each day.

- If you can incorporate the presence or absence of time cues (such as daylight or traffic noise — ones over which you have little control) into your daily routine, then so much the better.

- If you wake during the night (sometimes because you need to empty your bladder) then **do** return to bed and relax as soon as possible. **Do not** get up and make a cup of tea or a snack, as this will give misleading information about your habits to your body clock.

# Chapter 8

## Your body clock — some more details

We have discussed the evidence for a body clock and for the kinds of rhythmic change that it normally produces. What we have not yet considered is where the body clock is, how it works, and how it is adjusted by external time cues. These have been areas of very active research in the last decade or so, and many complex details are now known about the processes involved. Here, we will outline some of the current views, necessarily having to leave out many details and sources of controversy.

## 8.1 Where is the body clock?

We do not know the exact site of the body clock in humans, for the obvious reason that experiments to determine this are unethical. We do, however, have good evidence to indicate its whereabouts in several mammals, including those that are near-neighbours in an evolutionary sense.

Interest has centred on the role of two small groups of cells at the base of the brain in the hypothalamus, one on either side of the brain. They are called the suprachiasmatic nuclei (SCN). Evidence that the SCN make up the body clock can be summarized as follows:

1. Rhythmic electrical activity can be recorded from the nerve cells that make up the SCN and this rhythm has a period of about 24 hours. Clearly this is a promising start for a clock, but it might indicate that, instead of being the clock itself, these cells merely receive a rhythmic input from another region of the brain which is the clock. By analogy, nerve cells in the spinal cord show activity whenever a particular movement is made by the arm. These cells are not, however, the origin of arm movement but, rather, are driven by those regions of the brain that are.

2. Removing the SCN abolishes the rhythms of feeding, drinking, body temperature, and activity. Although this result would be expected if the SCN were the clock, as considered in (1) above, there would be the same result if the SCN were part of the transmission pathway by which the body clock communicates with the rest of the body.

3. Neural connections between the SCN and most other regions of the brain have been cut, and yet the SCN have continued to be rhythmic. This appears to remove the objection raised in (1) above, except that:

   • severing all the inputs to the SCN is not possible technically;
   • chemical links between the SCN and the rest of the brain are not removed by this technique, and so are still present.

4. Slices of the brain have been removed and incubated in special tissue-culture chambers. When these chambers have been maintained in constant conditions, the electrical and secretory activities from slices of the SCN, but not from other regions of the brain, have continued to show a daily rhythm. More recently, this result has been found when individual cells from the SCN, but not from any other region of the brain, have been cultured.

5. It has been shown that transplants of brain tissue containing the SCN can be used to generate rhythmicity in an animal that has previously been made arrhythmic by removal of its SCN.

The results in (4) and (5) indicate that cells of the SCN are sufficient by themselves to produce daily rhythmicity, and that an individual cell can do so.

As far as humans are concerned, there is anatomical evidence for a group of cells in the same region of the brain as the SCN in other mammals. This strongly suggests, but does not prove, that the SCN is the site of the body clock in humans also.

Location of a clock in the hypothalamus is likely to influence particularly those other functions that originate from, or close to, this area of the brain. Relevant to this are the following points:

• The hypothalamus also controls body temperature, food and water intake, sexual drive, and the secretion of several hormones.

• The hypothalamus is close to those regions of the brain which are involved in controlling the autonomic nervous system (which controls breathing, heart rate, and blood pressure, for example), sleep, and alertness.

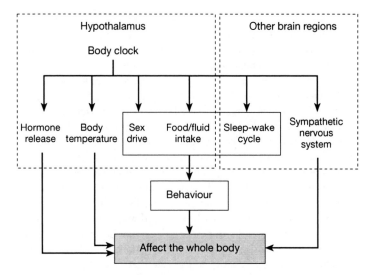

**Figure 8.1** Some outputs from the body clock.

Considering these two points together, it can be seen that the body clock can influence the body as a whole, not only by modifying behaviour but also by acting on variables — like hormones, the autonomic nervous system, and body temperature — whose effects permeate the entire body. Some of these points are summarized in Fig. 8.1.

## 8.2 What makes the body clock tick?

The information about the clock resides as clock genes in the nucleus, as part of the cell's DNA (deoxyribonucleic acid), its genetic information. Genetic information is coded as a specific sequence of molecules, called bases, on the DNA molecule. The first such clock gene to be discovered was found in the fruit fly and named 'per' (short for 'period'); another gene was then discovered and called 'tim' (short for 'time'). These genes, or ones very similar, have also been found in algae, fungi, and several animal species — an example of 'evolutionary conservation'. This suggests that, once evolved, the body clock has a strong survival value (see 'Some thoughts on Part I', after this Chapter).

Information from the clock genes is treated in the same way as is genetic information in general. The specific sequence of bases in DNA is first converted into a complementary sequence of bases in a molecule called messenger RNA (ribonucleic acid). This molecule then passes out of the nucleus into the cytoplasm. Here, the sequence is 'read' by organelles called ribosomes, which convert the sequence of bases into a specific sequence of amino acids (using the 'genetic code', which translates sets of three messenger RNA bases into specific amino acids). Thus, by this method, the specific information contained in the gene appears as a specific sequence of amino acids. This sequence of amino acids then folds up to become a functional protein, not surprisingly called a 'clock protein'. In other words, activity of the clock genes results in an increasing concentration of clock proteins in the cytoplasm of the cell.

All the roles of the clock proteins are not yet known, but they must include responsibility for producing, in some way, the electrical and secretory activity that can be measured in cells of the SCN. Next, the clock proteins interact with each other and with other substances in the cell to produce a series of products. These products then pass back into the nucleus, where they inhibit the actions of the clock genes. Therefore, messenger RNA from the clock genes, and the clock proteins themselves, are no longer produced, and so the concentrations of clock proteins and the products derived from them begin to fall. In turn, this means that the inhibition of activity of the clock genes (due to the clock protein products) is lost — and so the clock genes become re-activated, and the whole cycle starts again. This whole cycle takes about 24 hours, and it is the origin of the free-running, circadian rhythm of the cells of the SCN. It is illustrated in Fig. 8.2.

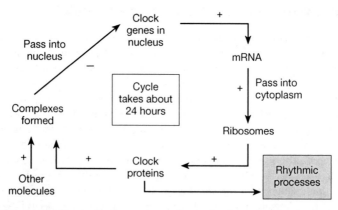

**Figure 8.2** The molecular and genetic mechanism of the body clock.

There are 'clock mutants' — animals or plants having a change in the properties of the clock genes or in the speed at which the clock proteins interact with other substances (see Fig. 8.2). This results in the period of free-running rhythms in these organisms being much shorter or longer than the norm. For example, mutants with free-running periods of about 20 hours or 29 hours are known. These mutants have been produced in the laboratory, but are not found in animals living in the wild. Presumably, the possession of such a body clock — useless for adjusting to the solar or tidal day — would place the individual at a lethal disadvantage.

## 8.3 How do time cues exert their effects upon the body clock?

As their name indicates, the cells of the SCN are situated immediately above the optic chiasm, an area where the two nerves which carry visual information from the eyes cross over each other on their way into the brain. Considering the importance of the light–dark cycle as a zeitgeber, such a position is, intuitively, appropriate, since we must expect that there will be some link between the eyes and the body clock. Indeed, it has been shown in mammalian species that there is a nerve, the retinohypothalamic tract, that runs directly from each eye to one of the SCN. This nerve branches from the optic nerves when the latter enter the brain, the main part of the nerve going to those regions of the brain that are involved in the sensation of vision. This suggests that the sense of vision is separate from the process by which the light–dark cycle acts as a zeitgeber and adjusts the body clock. The observation that some blind people have body rhythms that are adjusted to the normal light–dark cycle indicates that the pathway to the body clock continues to be functional in them, and supports this suggestion.

The detailed means by which the input from the eye adjusts the body clock is not yet clear. The first issue is that of deciding how light signals are picked up by the retina. The obvious possibilities are the rods and/or cones (the receptors in the retina that pick up the information required for vision). However, the body clock can be adjusted to the light–dark cycle in mice that lack rods and cones in their retina. What other receptors there might be is not clear.

There is not agreement on two related issues. First, are these receptors distributed evenly over the whole retina or, as with the cones that are responsible for colour vision, are they distributed in just part of it? If it is only part of the retina where they are found, which part(s)? Second, whereas the visual pigments (complex molecules that pick up the light signals) for rods and cones are based upon vitamin A (which is why eating carrots — a rich source of this vitamin — is said to be good for vision), it seems as though some other type of pigment, based on a molecule different from vitamin A, might be important for adjusting the body clock.

Returning to the problem of how the light–dark cycle might adjust the body clock, light exposure late in the evening and in the first half of sleep (that is, in the hours before the temperature minimum in humans and other diurnal creatures) causes a phase delay of the body clock. By contrast, light in the early morning or the second half of sleep (that is, in the hours after the temperature minimum) advances the body clock; and light at all other times has little or no effect. This relationship between the time of light exposure and the phase change (change in timing) in the body clock is known as the 'phase response curve'. The importance of such a curve is as follows. If the clock is running slow, and this is its natural tendency (see Fig. 2.2), then the body temperature minimum will be reached just before waking up. After waking up, therefore, light — particularly the bright light of outdoors, but domestic lighting also to a lesser extent — will advance the body clock. Conversely, if the body clock is running ahead of schedule, then the temperature minimum will be approached while the person is still awake — so light in the evening will delay it.

The phase response curve to light enables the body clock to be adjusted to run in synchrony with our light–dark schedule.

At a molecular level, it appears that the light signals pass to the SCN and produce changes that, ultimately, alter the cyclic activity of the clock genes and cyclic production of clock proteins. Consider the case where the light causes a fall in activity of the clock genes. If it occurs at a time when the activity of the genes is rising, then it will cause the process to have to start all over again; this has the effect of delaying the cycle and the phase of the clock. If it occurs when gene activity is decreasing (due to inhibition by the clock protein products), then this will have the effect of speeding up this process of gene inhibition and advancing the clock. If it occurs when the genes are inactive (after inhibition and before starting the next cycle), then there will be no effect upon the phase of the clock. A similar result can be achieved (an advance, delay, or no change in the timing of the body clock) if the effect of light is not upon the clock genes but rather upon the interactions between the clock proteins in the cytoplasm, altering the amount of suppression of gene activity that they will cause.

Therefore, even though details of the above processes remain to be elucidated, the general principles are known. These become important later, when adjustment to a new time zone (Chapter 11) or to night work (Chapter 13) or to certain clinical disorders (Chapter 9) are considered.

## What about other zeitgebers?

Melatonin is secreted into the bloodstream from the pineal gland at night (see Fig. 3.4). It, too, adjusts the body clock, but its phase-shifting effects are the opposite to those of light, since melatonin in the morning delays the body clock, and advances it in the evening. There are receptors for melatonin in the SCN, and it is through these that melatonin acts as a zeitgeber. The secretion of melatonin is suppressed by light, particularly bright light, and this means that light and melatonin normally act together to adjust the body clock. Thus, bright light on waking in the morning will advance the body clock, not only by the direct phase-advancing effect of light at this time, but also by suppressing melatonin secretion and so removing the phase-delaying effect that melatonin would exert at this time.

The possible role of meal times in adjusting the body clock has not been established in humans, although it has been suggested that a high-protein meal causes an increase in alertness, and that a high-carbohydrate meal promotes sleepiness. Without doubt, such meals will affect the relative concentrations of some amino acids in the bloodstream in opposite ways, and this change will alter their uptake into the brain. The high-protein meal will promote brain uptake of the amino acid, tyrosine, which is built up in the brain into noradrenaline, a substance that is released in the brain when we are active. By contrast, the high-carbohydrate meal will promote brain uptake of the amino acid, tryptophan, which is built up into serotonin and melatonin, both of which promote sleep when they are released in the brain. However, it has not been shown that the **release** of either noradrenaline or serotonin is increased by loading the brain with the appropriate amino acid. In other words, meals *per se* have not been shown to act as a zeitgeber.

When the possible role of activity as a zeitgeber is considered, it has been shown that introducing a hamster to a new running wheel at different times of the day can adjust the timing of the body clock. The direction of adjustment depends upon the time the running wheel has been introduced to the animal. By contrast, making hamsters run to keep warm (by placing them in a colder environment) seems to be less effective. That is, it seems to be the excitement associated with activity rather than the activity itself or the heat that activity produces that causes the phase adjustment — hamsters find a new running wheel very interesting!

Exercise in humans, however, seems to be less effective. Whether this is because we have not exercised long or hard enough, or do not find exercise sufficiently exciting, or are not as sensitive as hamsters to the effects of exercise, is not known. Certainly, our own studies have shown that even when sports science students (who enjoy exercise!) exercise for half an hour at a vigorous rate, it has no effect on the human clock. Nevertheless, there are input pathways to the SCN through which information about activity could pass.

In summary, there are pathways by which several zeitgebers might exert their effects. It seems to be the pathways transmitting information about the light–dark cycle, both directly and indirectly via melatonin release, that are, in practice, important in humans. These pathways are summarized in Fig. 8.3.

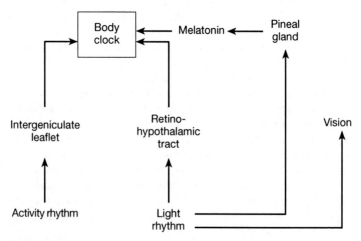

**Figure 8.3** Pathways by which light, melatonin, and activity rhythms act as zeitgebers.

# Now try this: a scientific problem (I)

To get some idea of the difficulties involved in studying the body clock — a structure hidden inside the brain whose activity can only be inferred indirectly by measuring the rhythms it is believed to produce — consider the following analogy.

A family of four is believed to be involved in organizing a spy-ring from their home. You have to prove this. You have free access to the family members outside their house, but when they are inside you can investigate them only by looking through the windows of their house. You have no helpers and can only devote a total of 4 hours' observation time to the task on any day. What is the best way to approach the task?

Your answer will depend upon your definition of 'best', of course, but some of the decisions you will have to make are listed below. (Comments in brackets refer to comparable difficulties when studying the body clock.)

1. When should you study the family? (You are rarely able to sample rhythms as often as you would like.)

   (a) Are your 4 hours of observations made in a single session or a number of shorter sessions?

   (b) If it is a single session, is this at the same time, or at different times, each day?

   (c) If it is at different times, then how do you arrange them?

   (d) If it is a number of sessions, then how many? Having decided upon this you then have the problems in (c).

2. How should you study the family when it is at home? (Do you concentrate on a large or small area of the brain? If the latter, which one, because other regions of the brain apart from the SCN — the pineal gland, for example — might be important?)

   (a) Do you follow one person only throughout your observation time(s) or choose, instead, a single place to see which of the family, if any, are there?

   (b) If you choose to observe a single place, then do you study a place that is large or small — the front garden, a single room, or the exit at the end of the drive?

   (c) If it is a single room, which one? Do you change the room? If so, how often? To which other place?

You might argue that such an investigation, though time-consuming, would enable you to build up the picture you want. But consider how the following additional constraints would make your task so much more difficult (and relate these constraints to the problems of being unable to study the SCN directly, and of being unsure of exactly how it then sends its information to produce the rhythms in, say, body temperature or urine flow — things that we can measure).

1. Not all members of the family belong to the spy-ring (not all cells in the SCN might participate in clock activity)

2. You are not quite sure exactly how the spies would communicate with others who might be involved. Do you investigate letters, phone calls, radio transmissions, or e-mails?

3. The house is surrounded by a high wall. You know that there are many paths leading to and from the house but:

    (a) You do not know how many there are, where they come from, or where they lead.

    (b) You cannot study some of these paths closely — they are guarded and can be viewed only from a distance.

    (c) In addition, there are so many visitors that you have no time to question them all — which would you choose?

It is not quite so simple now, is it? Even though you might continue to build up a picture of the whole operation, it must be based upon an increasing number of assumptions — and assumptions might be wrong, with potentially catastrophic effects upon the correctness of your account of the family and its activities.

In the same way, research experiments are continually testing incorrect ideas and going down blind alleys. But, slowly and painstakingly, the total picture, though initially distorted, begins to emerge more clearly — and it is this gradual piecing together of the jigsaw that makes research so interesting and rewarding.

# Some thoughts on Part I: the usefulness of the body clock

What advantage do animals and plants gain by possessing a body clock, since it can be argued that they respond to the rhythms in their environment anyway? (That, after all, is what is meant by the external cause of a daily rhythm.) What would be the disadvantage if there were no clock and a plant moved its leaves only in response to the sun? Or if the shore-dwelling creature left its burrow in the mud to forage only in response to the tide coming in? Or if the diurnal or nocturnal animal's response to sunrise and sunset were only a direct one? What is the advantage of a body clock to us humans, who can, anyway, respond to changes in posture and water content of the body and to changes in environmental temperature?

There must be some advantage, since all plants and animals so far studied possess a clock; once the clock had evolved, subsequent species kept it, with little further modification.

We suggest that there are two advantages to possessing a body clock. The first requires us to realize that, when an animal or plant responds to a change, **the response can take place only after the change has begun to exert its effect.** Particularly with plants, these responses might be rather slow. For example, the process of moving leaves into the right position to catch most sunlight takes time, so that, because the sun appears to move continuously, the position of the leaves will always be behind the ideal one. For an animal in its mud burrow, valuable time will be lost if it cannot prepare itself for foraging until **after** the tide has come in.

The advantage of a body clock is that **it enables its possessor to predict a regular change and so be ready for it when it takes place.** Thus the plant can, during the later part of the night, move its leaves to a position appropriate for sunrise; predatory animals can prepare themselves for the activity of hunting; hunted animals, and creatures that must not dry out when the tide recedes, can escape to safety before it is too late. For humans, as we have discussed in Chapter 1, the body has been preparing us for waking up since about 05:00 in the morning, so that, by the time we actually do so, we are better prepared for the rigours of a new day. In the evening, our body clock begins to 'tone us down' in preparation for sleep.

Responses to the environment are still required, of course, because the body clock is not a perfect timekeeper and, anyway, environments cannot be predicted precisely — the tide can be higher, food abnormally scarce, or the day more cloudy and thus darkness, earlier. Both mechanisms — that due to the clock and that due to the environment — are required by a living organism, and our previous discussion of the roles of the environment, as an external cause of a rhythm as well as a zeitgeber giving time cues to adjust the body clock, shows how intimate their combination is under normal circumstances. This inter-relationship between the environment and the organism is a fine example of the evolutionary response of a living organism to ecological factors. Time and rhythms are integral parts of the environment, and humans are constrained by them as much as other living organisms are.

The second advantage of the body clock is less easy to see, but no less important even so. The body clock allows different physiological and biochemical functions to be separated in time. Humans are one of the few species that takes only one consolidated sleep each 24 hours. During this sleep, growth, repair, and recuperation take place and, during our extended activity period, we can expend much physical and mental effort, as well as take in food and eliminate unwanted materials. To achieve this, it is necessary to ensure that the requirements for activity or rest are available at the appropriate time. The previous chapters have shown how this requirement impinges upon all aspects of our physiology and biochemistry. Humans work according to a 'body timetable' that is controlled by the body clock. Some aspects of this timetable are shown in the table on the next page. (Some of the items shown in this table will be covered in Part II of the book.)

A daily timetable for humans

| | Change | Effect |
|---|---|---|
| 20:00–24:00 (evening) | Falling temperature; falling adrenaline; rising melatonin | Falling alertness; beginning to prepare for sleep |
| | Falling cortisol | Airway constriction and increasing risk of asthma attacks |
| 24:00–04:00 (early night) | Greatest growth hormone release | Deep (SWS) sleep; fat metabolism |
| 04:00–08:00 (late night) | Rising temperature; rising adrenaline; falling melatonin; rising cortisol | Dream (REM) sleep; preparing for waking |
| 08:00 (on waking) | | 'Sleep inertia' |
| | Rising BP; rising activity; platelets stickiest | Greatest load on the heart and greatest risk of cardiovascular problems |
| 09:00–12:00 (morning) | Rising body temperature with little fatigue | Best performance at complex mental tasks |
| | Rising insulin sensitivity | Glucose metabolism |
| 12:00–14:00 (early afternoon) | Transient fall in adrenaline | Time for a siesta? |
| 15:00–20:00 (late afternoon) | Highest temperature | Physical performance best |
| | Effects of mental fatigue begin to show | Complex tasks deteriorate |

# Part II
# Your body clock in disorder

We have considered so far the daily rhythms of healthy humans. In Part II, we describe the effects of abnormalities of daily rhythms. These abnormalities arise either because there is an alteration in some aspect of the timing system itself, or because modern society has made demands upon us (long-haul flights and the '24-hour society') that have outstripped our biological evolution. But to end our book on the difficulties associated with jet lag and shift work would be too negative. So, we will end by describing an area in which the knowledge of daily rhythms has enabled a small but important contribution to medicine to be made — in the diagnosis and treatment of illness and disease.

# Chapter 9

## Abnormalities of daily rhythms

Put simply, the daily timing system consists of a body clock that is synchronized by external time cues and that sends information to a variety of systems in the body, so producing daily rhythms in them. It would not be surprising, however, if abnormalities were to arise at many points in such a system. Some of these are described below, together with ways to deal with them. A fairly extreme example is illustrated by a consideration of patients who are severely ill.

## 9.1 Patients in intensive care

Daily rhythms are abnormal in patients who have recently undergone major surgery or who are in an intensive care unit because of serious illness. There are several reasons why such abnormalities might arise.

- The patient, particularly one in an intensive care unit (which has artificial rather than natural lighting), is in an environment that is poor with respect to natural time cues.

- The patient's treatment — including artificial feeding, ventilation, and so on — is likely to be continuous rather than rhythmic. Drugs will be given if and when needed, probably with no obvious daily rhythm.

- The patient's perception of, and responses to, their environment are likely to be grossly reduced — they might be unconscious or paralysed, for example.

Thus, patients who spend long periods of time in an intensive care unit often report a form of disorientation — not knowing what time of day it is or where they are. The abnormalities of the timing system in such circumstances might take one of several forms.

There could be:

- A complete loss of rhythmicity (because the expression of the body clock, or even the clock itself, has been suppressed);
- A free-running rhythm with a period greater than 24 hours (because the time cues were too weak or the subject was unable to receive or respond to them adequately);
- A rhythm of reduced amplitude (because the external cause of a rhythm was reduced and/or because the internal cause was weak);
- An abnormal clock period (due to molecular changes in the body produced by the illness or its treatment).

In any particular case, the correct explanation is difficult to ascertain, and it is unlikely to be a priority for the clinicians to do so. It might be advantageous to promote 24-hour rhythms in the patient (if the medical circumstances permitted it). This might be achieved by increasing the strength of external time cues or by administering sedation and intravenous feeding on a rhythmic schedule. Its basis would have some parallel with the imposition of a daily routine and 24-hour environmental cues upon premature babies (see Chapter 7).

## 9.2 Daily phase changes in normal subjects

In normal health, the daily rhythms are generally far more stable, but that does not mean that there is no day-by-day variability. As we have seen already, individuals differ slightly in the timing of their rhythms, particularly if their chronotype is that of a 'lark' or 'owl' (Chapter 1). Even so, if an individual (whether a 'lark', 'owl', or 'intermediate' type) is studied for a number of days, particularly if the hours of work and sleep are regular, then the timing of the daily rhythms rarely changes by more than one hour from one day to the next. This degree of stability is normally found in the population as a whole.

By contrast, for those who are less bound by the demands of a regular lifestyle — those working or studying by themselves, for example — the rhythms can become less regular (see Fig. 2.4). This is a predictable result of having less regular zeitgebers and external components of the body rhythms. Nothing is wrong in these cases. Also, as we have already discussed (Chapter 7), daily rhythms became less marked in ageing subjects. This change can reflect a decreasing output from a deteriorating body clock and/or some decrease in the dichotomy between activity in the daytime and consolidated sleep during the night. These factors, together with the possibility of a decreased strength of zeitgebers or a decreasing ability to respond to them, result in the timing of daily rhythms showing a greater day-by-day variation than in younger subjects.

## 9.3 Abnormal timing of daily rhythms

For elderly subjects who suffer from various forms of senile dementia, the timing of daily rhythms can show a far wider variation, at least when the sleep–wake cycle is considered. Times of sleep and activity can vary considerably from day to day, the sleep–wake cycle even becoming temporarily inverted on some occasions (Fig. 9.1, top). This causes consternation to other patients and to the patient's care-givers. There might be several causes of this abnormality, but a general deterioration of brain function, including that of the SCN (its input and outputs), is the prime suspect.

Most of us experience 'Monday morning blues' — the feeling of tiredness and low spirits on the first day back at work after the weekend of rest. In part, it is because our body clock has adjusted readily enough to the delay in our lifestyle at the weekend (remember that the free-running period is more than 24 hours), but it is less easily advanced on Sunday night and Monday morning to a timing more appropriate for work (see Chapter 2). Most of us are back to normal by Tuesday, however, having achieved this advance. This is a natural consequence of our lifestyle and shows that changes in the timing of our body rhythms take place in response to changes of our habits.

By contrast, some individuals show a timing of their rhythms that is fairly constant from day to day (that is, they can adjust the body clock to a 24-hour day), but it is always very delayed; they might wish to sleep from about 04:00–12:00, for example (Fig. 9.1, middle). These individuals appear to act like extremely pronounced 'owls' and are said to suffer from delayed sleep phase syndrome (DPSP). Importantly, if they are allowed to do so — as would be the case at the weekends and on holiday, for example — then the quantity and quality of sleep all seem to be normal. Moreover, the timing of the body temperature and plasma melatonin rhythms is normal, *but with respect to the delayed sleep–wake cycle*. In other words, the whole circadian system is timed much later than in the normal population.

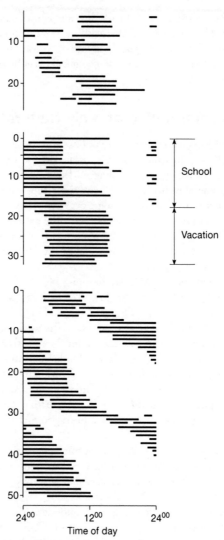

**Figure 9.1** The sleep–wake records of: (top) an elderly patient suffering from senile dementia; (middle) a young subject suffering from DSPS; (bottom) a young man with normal sight but suffering from periodic insomnia due to the presence of a 'free-running' rhythm. In all cases, the times of sleep are shown by a bar and successive days are shown below each other.

Advanced sleep phase syndrome (ASPS) is rarer, but it is similar, except that the subjects can be considered to be pronounced 'larks'. They wish to sleep from about 18:00–02:00, for example, with all their rhythms advanced by the same amount.

Both disorders are incompatible with a conventional lifestyle, though it might be noted in passing that DSPS and ASPS sufferers might be better suited than normal individuals to working night shifts and morning shifts, respectively. The cause of the syndromes is unknown but, again, suspicion falls on some abnormality of the body clock itself, or of the way in which it is adjusted by zeitgebers. Similar abnormalities have been found in laboratory-bred populations of fruit flies, and so a genetic contribution is possible. It can be speculated that this condition might lead, over an evolutionary time-scale, to the development of a group of 'nocturnal' rather than 'diurnal' individuals. Such a separation might even form the basis for evolution of a new species.

## 9.4 Inability to adjust to a 24-hour day — 'free-running' rhythms

There are some individuals who, even when attempting to live normally, do not remain in synchrony with the 24-hour day. Instead, they 'free-run' with a period of about 25 hours (Fig. 9.1, bottom). One example was of an individual who enjoyed a good job with normal daytime hours. His body clock free-ran so that on occasions he was the victim of a clash between an internal cause (a body clock, which acted as though it was night and tried to promote sleep) and an external cause (society, which required him to work in the daytime). Under these circumstances, he would feel tired at work and yet have difficulty in sleeping at night. At other times, the demands of his body clock and society coincided, but then much of the time would be spent catching up on lost sleep. In fact this individual found the stress of such an existence too much and was forced to leave his job.

For those without a job or social commitments that tie them to a 24-hour day, this disorder need not be a disadvantage. A single, intro-verted individual, capable of catering for him or herself, and whose job did not require person-to-person contact with others (e-mail would be sufficient, for example), would be able to live like this. Such a person could be cocooned in an artificial world in which meals and lighting were adjusted to accord with the dictates of the body clock. This would be like the subject in a cave (in Chapter 2) and the person's sleep pattern might look something like that shown in Fig. 2.2.

This form of insomnia — periodic insomnia — is rare in individuals who possess the faculty of sight; but, because the light–dark cycle is such an important zeitgeber in humans, this type of insomnia is, not surprisingly, more common in blind people.

## 9.5 Advice to those with sleep problems due to the body clock

The cause of the above abnormalities is rarely known, but it seems to be some combination of the following:

- Zeitgebers — they are too weak, not sensed sufficiently well, or transmitted inadequately to the body clock.

- Body clock — this is insensitive to zeitgeber inputs, or it possesses an abnormally long or short free-running period.

- Output from the clock — clock information about phase is not transmitted adequately to the centres controlling body temperature, hormone release, or the sleep–wake cycle.

Whatever the cause(s) might be, the main problems with body clock abnormalities are the inability to sleep soundly at night and the feeling of fatigue during the daytime. Undoubtedly, other rhythms will be similarly affected, but these abnormalities are much less likely to be noticed by individuals, unless they indirectly affect sleep (as in the case of an abnormally high rate of urine production at night, that would force a person to wake up in order to empty their bladder).

Even accepting that sleep problems are the main ones, the use of sleeping pills to promote sleep cannot be recommended, if only because they are not attacking the cause of the problem. An approach which has been used successfully, and which attempts to deal with the body's timing system, has been to strengthen the zeitgebers of the individual, to increase the individual's awareness of them, and to increase the regularity of the individual's lifestyle — approaches that will often overlap very considerably. These problems have been considered before (see the 'Now try this' sections at the end of Chapters 2 and 7), and aged subjects who lead successful lives often have a very regular lifestyle. Moreover, the problems will arise again when advice on jet lag (Chapter 11) and shift work (Chapter 13) is given.

Any individuals who have difficulty in adjusting to a 24-hour day should make every attempt to strengthen rhythmic factors and adhere as rigidly as possible to a regular routine. Strong attempts should be made to sleep only at night — to forego 'forty winks' after lunch or dinner, for example — and to be occupied during the daytime. Taking a shower or some gentle exercise on waking (procedures that are aimed at stressing a regular start to the day) should be considered. For older people, finding, or being given, something to do need not involve tasks that are particularly demanding, either mentally or physically. It might be sufficient just to browse through a photograph album — reminiscing is a good way to get a group of people involved. A mealtime also can become more of a social event.

The external environment can promote stability of the body clock. In the daytime, lighting should be bright, particularly during a regular part of the day. The general atmosphere could also be enlivened by background music. At night, the opposite should hold, so that preparations for sleep become quite natural.

These methods have been found to be valuable in many clinical situations, particularly with older patients suffering from senile dementia.

Another method — the regular intake of melatonin in the evening — can also be useful. Since the pineal gland starts to secrete melatonin into the bloodstream during the evening, and this promotes a fall in body temperature, the hormone seems to act in humans to promote sleep. Melatonin might also act as an 'internal zeitgeber', and it is in this role that it has been found to be useful in blind people and for others who show free-running rhythms. Its use adjusts the free-running rhythm to one with a period of 24 hours, and there is a decrease in the symptoms of periodic insomnia.

In the case of DSPS and ASPS, the treatment is simple, and can be seen as a special example of changing one's lifestyle to become less of an 'owl' or 'lark' (see Chapter 2).

For DSPS patients:

1. They are asked to go to bed 3 hours later each 'night' and to get up 3 hours later each 'morning' — they will find this easy to do.

2. After several days on this schedule the patients will be going to bed and getting up at their chosen time, one that gives them enough sleep during the weekdays when they wish to work.

3. They must now stick rigidly to a 24-hour lifestyle — no late nights, or they will have to begin treatment all over again!

ASPS patients are asked to advance bedtime by 3 h each 'night' until they are sleeping at a convenient time. Then they must be warned against getting up or going to bed too early.

The above treatments, though effective, cannot be regarded as cures, since the underlying cause of the problem, even if known, has not been tackled. They are effective in the same way as a hearing aid is effective — by exaggerating the stimuli to which the underperforming system, be it the body clock or the ear, normally responds. There is also the advantage that the treatments are cheap and do not involve the long-term use of drugs.

## 9.6 Seasonal Affective Disorder

Strictly speaking, a consideration of Seasonal Affective Disorder (SAD) is outside the remit of this book, since it is a seasonal or annual rhythm. Nevertheless, it is included since it is believed to involve a disorder of daily rhythms, and because it illustrates some of the difficulties of interpretation that can arise in rationalizing treatment.

Seasonal Affective Disorder is associated with depression that is marked during the winter and early spring, and regresses in the summer. It has been suggested that it occurs when susceptible individuals do not get enough natural light during the winter months. In support of this view is the finding that the incidence of SAD amongst different populations increases as one passes from the equator towards higher latitudes — as do the length of the winter nights and the likelihood of waking up in the dark on winter mornings. It is unlikely that it can be attributed to cultural differences, since it has also been found when the incidence in different States of the USA has been investigated.

Early successes in the treatment of this disorder consisted of exposing patients to bright light in the dark winter mornings, mimicking the longer hours of daylight of summer. However, since bright light in the dark winter evenings appeared to be less effective, and bright light given in the middle of the day was almost as effective as when given in the morning, the explanation of the role of the light cannot be due simply to opposing the seasonal changes in the duration of natural lighting. Moreover, there has not been agreement as to whether there is a change in the timing or the amplitude of daily rhythms that is consistently found in sufferers from this disorder; some reports describe an advance in the timing of daily rhythms, others a delay, and yet others conclude that it is an increased variability in timing or a decreased amplitude of the rhythms.

Many reports of the effectiveness of light treatment include evidence that the symptoms improve at about the same time as the timing or amplitude of body rhythms becomes more 'normal' — though from the above, it can be seen that 'normal' might mean a delay in their timing, an advance, or a stabilization, or an increase in rhythm amplitude. Based on this mass of inconclusive evidence, some have concluded that there are several forms of the disorder, each associated with a particular set of rhythm changes.

That is, whilst the efficacy of light treatment is generally accepted, its rationale is still unclear — though, of course, this does not detract from the value of the treatment.

From a scientific viewpoint, there are some further issues that are raised by this consideration of SAD:

1. To test the hypothesis that SAD is due to a changed timing of daily rhythms, some patients should be given bright light when it would be predicted to worsen their symptoms, by adjusting their body clock in the wrong direction. Of course, there are ethical problems here but, in the absence of these control experiments, it might be that light is exerting its effect by some other more general means — say, by encouraging the patients to become more active, or by making them think that they will get better.

2. This last point raises the issue of the 'placebo effect'. By this is meant that a treatment might improve the patient's sense of well being (and so cause the symptoms of SAD to regress), not because of what the treatment is, but rather because the patient wants to recover, and so is responding to the clinician's attempts to help. Whilst such successful effects are to be welcomed, they do little to help an understanding of the underlying cause of SAD. There is another side to this. Based on what they learn from the media (from reading a book such as this, for example!), patients have certain expectations as to the type of treatment they should receive; this expectation can, in turn, influence the kind of treatment that the physician gives (or has to give).

3. It is quite possible that SAD is not primarily due to an abnormality of the body's timing system but is, instead, due to some other abnormality which causes several effects, amongst which are not only the feeling of depression but also changes to sleep, to activity, and to motivation. These can then affect the body clock indirectly.

In summary, it is not yet clear that a modification of the daily rhythms is a necessary requirement for treatment of SAD, or that an abnormality of the body clock is the cause of the disorder. It is possible that clinical cases are only an extreme form of a normal reaction of all of us to the shorter winter days. Many people are less excited about waking up and having to go to work in the dark rather than the light. Also, it might be that those who suffer from SAD differ from the majority who do not suffer, only because they are more susceptible than the rest of the population to feeling 'under the weather' when they do not see daylight, particularly in the morning.

If this interpretation is correct, then there are important implications for those who become subclinically depressed in winter when it is debated whether to change the clocks between the summer and winter, or rather to retain 'summer time' throughout the year. The extra hour of daylight in the winter evenings that retaining summer time gives is accompanied by an extra hour of darkness in the winter mornings.

There are other circumstances where the timing of the body clock is altered (for example, after a time-zone transition or during shift work), and which are associated with negative feelings. These rarely are severe enough to be classified as a 'depression'; they go under the description of a 'malaise'. However, they do affect different individuals to different extents. It is to these that we turn in the next chapters.

## Now try this: a scientific problem (II)

Consider the following comments about SAD. Can you devise ways of testing them? These are questions designed to test an understanding of the scientific method; *they are not a means of diagnosis or recipes for treatment.* If you think that you suffer from SAD, then you must consult a doctor.

1. In the Northern hemisphere, most SAD cases are referred to clinics in the months of December to April.
   (a) How could you test if the factor that causes SAD is a lack of light or the cold weather?
   (b) Would you get more information by considering patients who had suffered for several years rather than only one?
2. Bright light is often found to bring relief of the symptoms of SAD.

   How would you eliminate a placebo effect (that is, the treatment works because the patient expects, or hopes, that it will)?

3. Carbohydrate (sugar) intake is found to increase in patients with SAD.

> How could you show that this was not because chocolate bars, sweets, and so on, are more likely to be eaten, since the depression means that you cannot be bothered to cook a proper meal?

## Comments

1. In the Northern hemisphere, most SAD cases are referred to clinics in the months of December to April.

   (a) Ask patients when their symptoms first appeared. (This assumes they can recall this accurately.) Does this time tend to correspond to spells of cold weather, to times of dull weather, or to the length of daylight? (You will need to refer to local meteorological records to establish this). Does the onset of SAD symptoms depend upon whether the patient lives an 'outdoor' or an 'indoor' lifestyle? (It could be hypothesized that the outdoor type is more aware of the changing pattern of light.)

   (b) The patients could keep a diary of their moods throughout the year (a form of self-diagnosis). Moreover, they are likely to consult their doctor after a shorter delay if their SAD is a recurrent problem — a suggestion that could be tested statistically.

2. Bright light is often found to bring relief of the symptoms of SAD.

   Give some other 'treatment' that does not involve bright light.

   (a) Exercise at the time corresponding to light treatment? (How much exercise? What kind of exercise?)

   (b) Sitting down quietly for the same period of time that light treatment would have been given, but in front of a set of dim lights?

   (c) Giving noise or music instead of bright light? (What kind of music?)

   Unfortunately, patients may have learned from media coverage that light is somehow involved, so they will probably have a prejudice against other treatments. Giving bright light at a time that is hypothesized to worsen the condition is useful as a scientific test of the rationale underlying treatment, but it is ethically unacceptable, as has already been mentioned.

3. Carbohydrate (sugar) intake is found to increase in patients with SAD.

   (a) Make all types of food (protein, fat, and roughage, as well as carbohydrate) equally easy to obtain. Do they still prefer carbohydrate to other types of food, and is this the case whether or not they are suffering from SAD at the time?.

   (b) Ask patients to keep a daily record of the time it takes to prepare their food. Do they spend less time when they are suffering from SAD?

In practice, these investigations might require constant supervision of the patients over the course of several weeks, both in the winter (suffering from SAD) and in the summer (no symptoms of SAD). This procedure is time-consuming, and many patients would not cooperate, particularly in the summer when they are free of the symptoms of SAD. Even if they did cooperate, the problem that patients might choose the type of food because of habit, the weather, or what is (seasonally) available — rather than because of a careful consideration of their appetite — still exists. (In fact, there is evidence for carbohydrate *craving*, so it is not just a matter of apathy about cooking.)

Finally, there is the problem that it is assumed that the patient's habits are 'normal' when there are no symptoms of SAD, and that their habits change only when they are suffering from it. This need not be so; it might be that the abnormality (whatever its nature) is always present, but only manifest when the days are short. By analogy, consider hay fever sufferers; they do not suffer the symptoms of hay fever in winter (because the pollen that acts as the allergen is not being blown about by the wind), but the basic problem (hypersensitivity to certain natural substances) is still present.

We hope another point is made by these questions and comments — namely, that the problem is very complex and requires a great deal of meticulous study, and that the results are often difficult to interpret unambiguously.

# Chapter 10

## Long-haul flights and time-zone transitions — the problems

The tour operator for a long-haul flight to a distant land will stress the excitement of it all — the new places, people, and lifestyles, and the sense of romance and adventure. Not surprisingly, the discomfort, inconvenience, and disruption caused by such a flight, together with feeling 'below par' during the first few days after the flight, will not be discussed as much — if at all.

There are two separate problems associated with long-haul flights — 'travel fatigue' and 'jet lag'. The former refers to sheer tiredness, the latter to a disturbance of the body clock.

## 10.1 Travel fatigue

This condition is a combination of two factors. The first is the general hassle associated with the disruption to our routine caused by the travel, and the second is the discomfort of the flight itself.

The disruption to our routine includes the preparations for travel — arranging foreign money, passport, visa, tickets, as well as stopping the milk and papers, and getting somebody to look after any pets. On the day of travel itself, there is often an early start, queues at the check-in, and then further waiting around until it is time to board the plane. The journey might consist of more than one flight, in which case there will be more waiting around (or a hectic dash because of a late connection) and possible further complications in relation to luggage and immigration. After arrival at the final destination, there is, hopefully, the collection of luggage, and another round of passing through customs. Then it is time to find our way to the hotel, sometimes in a country where we have little or no understanding of the language.

These problems do not require the flight(s) to be long and arduous but, if they are — and this is likely in the case of intercontinental flights — then there is the added discomfort caused by the conditions in the flight cabin itself. Room can be very restricted and the air in the cabin is very dry (it has an oxygen pressure and water vapour content that are low in comparison with 'fresh air'). This dryness causes dehydration, which can lead to headaches, dry lips, and a dry nose. The lack of space can lead to muscle cramp, and there is evidence to suggest that it might also increase the incidence of deep vein thrombosis — 'economy class syndrome'. Deep vein thrombosis is due to the formation of a blood clot and might be produced by blood pooling in the lower limbs as the passenger stays inactive in their seat.

Even if things have gone without any hitches, the traveller will be tired, dehydrated, and irritable on arrival at the hotel. This is what is known as travel fatigue. We might be forgiven for being sceptical about the picture painted by the tour operator and the guide books!

## 10.2 Reducing travel fatigue

How can we reduce it? Some of the problems can be minimized by careful preparation — packing and getting together money, passport, and visas well before the day of the flight, for instance. It is advisable to leave plenty of time for catching the flight and for catching any subsequent flights, if the journey has to be broken — you can be delayed by traffic jams, or by your late arrival after the first leg of your journey.

### Drinks

Coffee, tea, and alcohol are likely to be offered, but try to limit how much of these you drink. This is not a 'killjoy" attitude, but an attempt to guard against the dehydration produced by the dry air in the cabin. All these drinks possess diuretic properties, which means they increase the loss of body fluids in the urine. Fruit juices, squashes, and spa waters are recommended alternatives.

### Food

This necessarily tends to be of the 'convenience' type and low in fibre, and might cause constipation later on. Try, instead, to eat foods that are higher in roughage; fresh fruit, wholemeal bread or rolls, salads, or carrots are all high in fibre. You might even take some fresh fruit with you for the journey.

## Exercise

Try to get some exercise during the flight to decrease stiffness and the possibilities of cramp and swollen ankles. For some, a stroll down the aisles is sufficient, but isometric exercises (alternate tensing and relaxation of muscles), or those involving only limited movement (of the neck, back, arms and legs, for instance) can be performed conveniently within the confines of your seat. Some airlines now promote this idea by showing an in-flight video that leads passengers through such a work-out. Many experienced passengers choose an aisle seat so they can stretch and leave their seat more easily. It is a good idea for tall passengers to book a seat by an emergency exit, where there is more room for stretching the legs and avoiding getting stiff. Such exercise should also guard against the possibility of deep vein thrombosis. Magazines and popular medical journals now recommend the wearing of special support stockings or socks on the flight — though how effective they are in this situation is not known.

## Whether to nap or stay awake

Boredom and fatigue will lead to the tendency to take naps but whether this should be encouraged or resisted depends upon details of your journey, and is discussed below. With some airlines and for first-class passengers, there is the facility (for example, the 'sleeperette') to take a good sleep rather than just a nap.

The good news is that travel fatigue wears off quite quickly. On arrival at your destination, take a shower, have several non-alcoholic drinks (to replace lost water), and start to relax. It is likely that you will be ready for bed by the time it is night, and so you will start the next day feeling refreshed — **unless** your journey has taken you across several time zones. In this case, you will have to deal with jet lag, and getting a good night's sleep might not be easy.

## 10.3 Jet lag

How does jet lag affect us? Why do we get it? Most importantly, what can we do about it?

The symptoms of jet lag are being experienced by more and more people as long-haul flights, spanning the globe, become more common. Symptoms constitute an ill-defined group that affect people in different ways and to different extents. Typically:

- A person suffering from jet lag will feel tired throughout much of the daytime, and yet be unable to sleep well at night. He or she will feel wide awake at bedtime and have difficulty getting to sleep, or will get to sleep but then wake up too early. This will increase the sense of general fatigue and might result in headaches and difficulties in concentrating on work and being motivated to do it.

- The person might know that it is time to eat — a grumbling stomach will make this clear — but appetite might be poor and any food that is eaten will cause a sensation of feeling over-full or bloated. Furthermore, it might not taste so nice, and indigestion might be more common.

- Bowel movements will often occur at inconvenient times in the night or at unaccustomed times during the day, and the traveller might suffer constipation or mild diarrhoea for a few days. Stools might be harder for a few days as waste material stays longer in the lower intestine and loses more of its water content.

In summary, the traveller feels 'below par'. The older person is often said to suffer more than the younger one because the older are more 'set in their ways', but there is also evidence that they are better able to 'pace themselves', in which case they suffer less. The symptoms are generally worse the more time zones that are crossed, and they are more marked after eastward flights than those to the west. Flights in a north–south direction (or vice versa) give less trouble — causing only travel fatigue, as described above.

## 10.4 What causes jet lag?

Why is the body so disorientated? It is a combination of two factors — the change in time zone and the sluggishness of our body clock to adjust to change.

### Time zones

The Earth revolves on its axis once every 24 hours. As a result, each country (except those close to the Poles) will experience sunrise, the passage of the sun across the sky, sunset and night once per revolution. Because of the direction in which the Earth revolves, the sun always rises in the east and sets in the west. As we travel to the east, therefore, these events occur earlier and, on a westward journey, they take place progressively later. The time at which the sun is highest in the sky is affected in the same way, this moment being defined as noon by local time. This has the obvious advantage that all countries match their local time to the hours of daylight, but it also means that different countries have different local times.

In order to standardize all these local times, the world has been divided into 24 time zones — each spanning 15° longitude. All time zones are related to one another: countries to the east are in time zones with local times that are ahead, whereas countries to the west have time zones and local times that are delayed. In practice, all times are related to a single one — local time in the 'prime meridian' (0° longitude) — which passes through Greenwich in England. It is called Greenwich Mean Time (GMT) or Universal Time (UT). Thus, when it is 16:00 in England, it is 11:00 in New York (5 time zones to the west), 08:00 in Los Angeles (8 time zones to the west), 20:00 in Abu Dhabi (4 time zones to the east), and midnight in Singapore (8 time zones to the east).

In principle, an identical calculation can be carried out using any time zone as the baseline, but the position is complicated in two ways. First, countries modify local time from the time zone they are in by up to 2 hours, according to whether it is winter or summer, in order to make more use of the daylight hours. Second, if travellers start from London and go eastwards halfway round the Earth, they will have passed through 12 time zones and gained 12 hours with respect to GMT. They will meet other travellers who have travelled westward and so are 12 time zones (12 hours) behind GMT. At this point on the Earth (180° longitude) the International Dateline is reached.

Suppose it is 12:00 (noon) on a Monday by GMT; at this moment, to the traveller at the International Dateline who has travelled eastwards, it will be midnight at the end of Monday; by contrast, for a traveller in the same place but who has travelled westwards, it will be midnight, but at the beginning of Monday. That is, crossing the International Dateline means that we either gain a day (eastwards crossing) or lose one (westwards crossing). This plays a key role in Jules Verne's novel *Around the World in 80 Days*.

A further consequence of having time zones is that when we travel we have to adjust to local time in the new time zone. It is this adjustment which causes difficulties for our body rhythms.

## The body clock and time-zone transitions

The body clock enables us to synchronize ourselves with our rhythmic environment (as was considered in more detail in the summary at the end of Part 1). An important property of such a clock is its stability. This means that its timing does not alter much from day to day, and transient changes to a normal sleep–wake cycle, whether these are a daytime nap or a brief nocturnal awakening, have no effect upon the timing of the body clock. Normally, these properties are advantageous (which is why they have evolved through natural selection), but they become disadvantageous after a time-zone transition, when a rapid adjustment of our body clock to the new local time is desirable.

For some days after the flight, the timing of the body clock — and with it the timing of our daily rhythms — will tend to be more appropriate for the time zone we have just left. Not only will we feel disorientated but also there will be more objective measurements to indicate that the flight has disorganized us. If we measure our rhythms in mental performance, for example, we will find that the normal daytime peak, with a morning rise and an evening fall, is absent. Instead, (Table 10.1) we might find that performance is improving right up to bedtime and deteriorating after sleep (eastward flight), or that it is best soon after waking and deteriorates throughout the day (westward flight). Superimposed upon such a rhythm will be a general decline in performance due to loss of sleep and fatigue. Moreover, daily rhythms in leg strength (Fig. 10.1), body temperature, and adrenaline will show similar abnormalities of timing.

'Jet lag' is the consequence of a mismatch between the body clock and external time, a mismatch that exists because the body clock is slow to adjust to the change in local time. Knowledge of this fact, together with an understanding of some of the properties of our body clock, will enable us to promote adjustment of our clock to the new time zone and so reduce the difficulties of jet lag.

## 10.5 Other explanations of jet lag

Our explanations of jet lag, however, should not go unquestioned. Accordingly, we offer some comments on alternatives that have been given. As is often the case, they contain some truth, and can be used constructively when attempting to deal with travel fatigue and jet lag.

**Table 10.1** Some differences from living in the home time zone experienced in the first days after flying east (East) or west (West) across 8 time zones

| Local time | Home time zone | East | West |
|---|---|---|---|
| 08:00–10:00 | Waking up | Ready to go asleep | Going well |
| 14:00–16:00 | Going well | Only just beginning to wake up | Ready to go to sleep |
| 20:00–22:00 | Preparing to wind down | Going well | Want to be asleep |
| 02:00–04:00 | Sleeping well | Still going well — not sleepy | Woken up and ready to go |

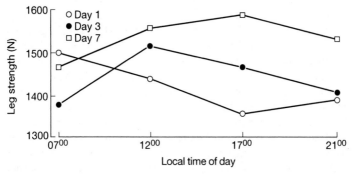

**Figure 10.1** Change in local timing of the daily rhythm of leg strength (measured in Newtons, N) after a westwards flight across 5 time zones. By day 7, the rhythm is adjusted to the new local time and shows a peak in the late afternoon. By contrast, the rhythm is unadjusted on day 1, showing a peak in the early morning, and still unadjusted on day 3 (peaking at about noon).

## Is it 'in the mind'?

This is partially true. Many of the symptoms of jet lag relate to how we feel, but such an explanation seems less acceptable when we would expect the mind to be helping us to enjoy ourselves on holiday, or to perform at our peak for business or athletics engagements, for instance. Moreover, the abnormal timings of bowel habits, of hunger, and, particularly, of body temperature, performance, and hormone rhythms are difficult to explain in these psychosomatic terms. Certainly, expecting that you will suffer from jet lag hardly helps, and this becomes an argument for adopting a positive attitude to the difficulties and knowing that they can be minimized by following the advice in the next chapter.

## Is it due to the inconvenience of intercontinental flight?

Disruption of our normal routine, loss of sleep, waiting for the flight to be called, and the apprehension associated with flying will all take their toll. To be sure, these general inconveniences of a flight disrupt our routine and make us tired, but this is travel fatigue, the effects of which wear off by the second day at our destination. The severity of jet lag depends upon the number of time zones crossed rather than the length of the flight and the hassle associated with it. This is why jet lag is not so marked after a flight in the north–south direction.

To illustrate this difference between travel fatigue and jet lag, consider laboratory-based experiments. The effect of time-zone changes, but without problems from the hassle of the flight itself, can be studied in experiments in which the subjects live a normal routine by the clock and then, suddenly, the clock is changed. For example, at 12:00 (noon) the experimenter might change the clock to 20:00, so mimicking a time-zone transition of 8 hours eastwards (Los Angeles to London, or London to Singapore, for example). The subjects must continue to live their normal routines, but now in accord with the new local time. Travel fatigue (the problems associated with the inconveniences of flight) has been removed, but has jet lag? In such simulated time-zone transition experiments, changes to the body's rhythms are observed to be very similar to those after real time-zone transitions. In addition, many components of jet lag (altered sleep, increased fatigue, a loss of appetite, changed bowel movements) are found, and the rate of adjustment of the body rhythms to a simulated time-zone transition is similar to that observed after real transitions.

Thus, many of the problems of time-zone transitions can be modelled in the laboratory where travel difficulties are no longer relevant. Even so, becoming angry when travel arrangements go wrong can hardly help. In brief, keep calm!

## Is it due to a change of lifestyle?

The 'change of lifestyle' argument is unsound because the change is not necessarily greater the further you fly or the more time zones you cross. Consider, for example, the cultural differences for a Western European journeying to parts of Central Africa (large cultural changes, but few time zones crossed) or New Zealand (smaller cultural changes, but more time zones crossed). The traveller to New Zealand has more jet lag to contend with than the traveller to Africa. Equally, the USA citizen travelling to the Brazilian rainforest has less jet lag (but a larger change in lifestyle) than one travelling to Western Europe. In addition, jet lag is experienced on returning home after a stay in another country, even when travellers are returning to their own culture.

Alterations in diet and even hygiene might well produce indigestion and 'traveller's tummy' but, again, this is determined by the food eaten rather than the number of time zones crossed. However, as is confirmed by laboratory experiments (where the diet can remain unchanged), time-zone transitions produce changes in appetite and bowel movements.

## Is it a matter of experience?

It could be argued that, with experience, the body becomes more able to adjust to a time-zone transition — a kind of 'learning' effect. Some frequent travellers (including aircrew) do, indeed, maintain that they get used to jet lag, but others find that the problem becomes worse. Travellers might get used to jet lag in the sense that they learn to live with it, or they might find it progressively more irksome as their initial excitement with travel begins to wear thin. The idea that the body clock might adjust more readily with practice is an attractive one, but there is no evidence to support it. What might arise with experience is the development of a better routine for coping with jet lag, and this will be one of the aims of the next chapter.

## 10.6 Adjusting the body clock

Accepting that the body clock is slow to adjust to a time-zone transition, our aim must be to speed up this adjustment. Adjustment will be speeded up if we can strengthen the time cues (zeitgebers) in the new time zone. We will just recapitulate what was described in Chapter 2.

The main zeitgebers are our rhythms of exposure to light and dark, and the secretion of the hormone melatonin. Rhythms in the sleep–wake cycle, in physical activity (including exercise, walking, gardening), mental and business activity (including appointments, shopping, office hours), social influences and leisure time (including the pub, disco, theatre, parties), and mealtimes, might all play a role too — if only to determine our exposure to the light–dark cycle. Under normal circumstances these different rhythms give the same information about external time; they reinforce one another. This can be seen when the differences between night (when it is dark and quiet, and we sleep) and day (a noisier and more 'active' environment when we arrange our meals and activities) are considered.

Such knowledge will be the basis for the advice we give in the next chapter for promoting adjustment of the body clock to the new time zone.

## 10.7 Check-list for travel fatigue

**Symptoms**

- Fatigue
- Disorientation
- Headaches
- 'Travel weariness'

**Causes**

- Disruption of normal routine
- Hassles associated with travel (checking in/baggage claim/customs clearance)
- Dehydration due to dry cabin air

**Advice**

- Before you go:
    - Plan the journey well in advance
    - Try to arrange for any stop-over to be comfortable
    - Check about documentation, inoculations, visas
    - Make arrangements at your destination
- On the plane:
    - Take some roughage (for example, apples) to eat
    - Drink plenty of water or fruit juice (rather than tea/coffee/ alcohol)

- On reaching your destination:
  - Relax with a non-alcoholic drink
  - Take a shower
  - Take a **brief** nap, if you wish

(Advice on jet lag is given in the next chapter)

# Chapter 11

# Time-zone transitions — advice

Advice on whether or not to adjust to the new time zone depends very much upon the kind of journey being undertaken. We will consider the following possibilities:

- A flight crossing only a few (3 or less) time zones;
- A flight crossing several time zones (to the west or east), when the return journey back home occurs after only a short (less than 4 days) time;
- A flight crossing several time zones to the west, where there is sufficient time for adjustment to the new time zone (staying 4 or more days in the new time zone);
- A flight as in the last example, but to the east.

## 11.1 How many time zones will you cross?

Table 11.1 gives a general guide to the number of time zones between countries on GMT (for example, the United Kingdom) and other countries. Corrections need to be made if single or double summer time is in operation – that is, local time is 1h or 2h, respectively, ahead of the time predicted for the time zone.

If the journey does not start or end in a country on GMT, then the list can be used by combining entries. Thus, for a journey from Saudi Arabia to the east coast of America:

- The east coast of America is 5 time zones to the west of the United Kingdom.
- Saudi Arabia is 3 time zones to the east of the United Kingdom.
- Therefore the journey is 8 time zones in a westward direction.

**Table 11.1** Time zones in hours to the west (W) or east (E) of GMT

| | Time-zone difference (hours) | Local time when noon GMT |
|---|---|---|
| New Zealand | 12 | Midnight |
| Australia | 8–10 E | 20:00–22:00 |
| Japan | 9 E | 21:00 |
| Singapore | 8 E | 20:00 |
| Delhi | $5^1/_2$ E | 17:30 |
| East Africa | 3 E | 15:00 |
| Saudi Arabia | 3 E | 15:00 |
| Moscow | 3 E | 15:00 |
| Central and South Africa | 2 E | 14:00 |
| Eastern Europe | 1 E | 13:00 |
| West Africa | No time difference | Noon |
| Greenland | 3 W | 09:00 |
| Brazil | 2–4 W | 08:00–10:00 |
| East coast of Canada | 3 W | 09:00 |
| East coast of USA | 5 W | 07:00 |
| West coast of South America | 5 W | 07:00 |
| Midwestern states of USA/Canada | 6–7 W | 05:00–06:00 |
| West coast of USA/Canada | 8–9 W | 03:00–04:00 |
| Alaska | 11 W | 01:00 |

If the country you are visiting is not listed in Table 11.1, you can find out how many time zones distant it is by the following means (but be warned that slight regional modifications are quite common, even though they are only of an hour or so).

• The telephone directory. The information can be found under the section on international dialling codes.

• A world atlas. Some give the time zones directly. If not, they can be estimated by knowing that one time zone equals 15° longitude. Thus, Christmas Island (105°E) is 7 time zones to the east (ahead) of GMT, and Jamaica (75°W) is 5 time zones to the west (behind) of it. Therefore, a flight from Christmas Island to Jamaica would be 12 times zones in a westward direction.

## 11.2 The flight itself

Much of the advice relating to the flight has been given in the previous chapter when travel fatigue was discussed. Arriving at your destination as fresh as possible is a good start, and the advice to combat travel fatigue should help in this.

## 11.3 Dealing with jet lag — some examples

The main issue now is to minimize the effects of any jet lag, once at your destination. As we shall see, your behaviour during the flight can begin the process either of adjusting the body clock to the new time zone or, if your journey indicates that this is appropriate, of retaining the timing of the time zone you have just left.

### Example 1: a flight crossing only a few time zones

If you are flying to the east and crossing only 1 or 2 time zones, or flying to the west and crossing 3 time zones or less, you are unlikely to suffer from the effects of jet lag. Even so, you are advised to leave at least one clear day between the flight and an important meeting. During this time you can overcome travel fatigue. Catch up on lost sleep (as long as your naps are not so long or frequent that they prevent a full night's sleep), relax after the stresses of the journey, replenish body fluid, adjust eating habits, and generally adapt to your new surroundings.

During the flight, you are advised to sleep or nap if it is night-time but, if it is daytime, try to read a book, talk, play cards, and so on, rather than nap. In addition, if the flight is long, take some exercise at intervals to combat the risk of deep vein thrombosis or a blood clot developing.

### Example 2: a flight crossing several time zones but only a short visit

If your stay in the new time zone is brief (less than 4 days), then a substantial adjustment of the body clock is unlikely. Therefore, you are advised to arrange important meetings at times coincident with daytime on your home time, and to avoid times coincident with night on home time. This strategy means that, after a flight to the east, meetings should be arranged to take place in the later half of the daytime rather than in the morning by new local time. By contrast, after a flight to the west, meetings should be arranged to take place in the morning by new local time rather than in the later part of the day.

If you feel tired, a short sleep before an important meeting can be beneficial, as long as it ends at least one hour before your meeting — to make sure that you have woken up fully. Sleeping pills are not advised unless they are short-acting benzodiazepines, whose effects wear off quite quickly, **and** you have previously found them to be useful, **and** have discussed the matter with your doctor.

## Flights crossing several time zones and staying at your destination for several days

If your stay in the new time zone is longer than 3 days, then you can attempt to promote adjustment of your body clock. This entails matching the timing of your lifestyle (times of sleep, activity, and meals) to that of the new time zone as fully and rapidly as convenient. This will have the effect of strengthening the zeitgebers that adjust the body clock. In addition, for the first few days in the new time zone, you should try to seek out and avoid bright light at certain times (see below).

In most cases, following this advice, particularly with regard to sleep, will initially be against the 'natural' dictates of the body clock — but that is exactly what jet lag is all about. Obeying the body clock in these cases will only prolong the difficulties associated with living in the new time zone, with the body clock remaining timed appropriately for the time zone you left.

Unfortunately, following the advice below will not remove jet lag, but it certainly should decrease its severity and duration.

### Example 3: a flight to the west

Flying westwards requires you to delay the timing of your body clock. This adjustment is comparatively easy, since the body clock naturally tends to run rather slowly (Chapter 2).

**Before the flight**   Can you begin to adjust the timing of your lifestyle to the new time zone in the days immediately before departure? One possibility is to go to bed one to two hours later than normal each night and to get up one to two hours later each morning. Of course, this might not always be possible. It is rarely useful to try and adjust fully to the time-zone transition before the journey, since this will interrupt your normal lifestyle too much.

**During the flight**   Think ahead! Set your watch immediately to agree with local time at your destination. During the flight, try to sleep if it is night-time at your destination and, when it is daytime there, try to stay awake (find somebody to talk to, read a book, or watch the in-flight movie). If sleep is appropriate, a facility allowing a good sleep (a 'sleeperette', for example) can be a great advantage over having to try to nap in your seat.

Be warned, however, that there are two practical problems that might arise if you attempt to adjust your sleep times in the way suggested. First, the in-flight schedule (including meals) is not always arranged in accord with local time at your destination. Second, if you have to break your journey or change flights, make sure you use the appropriate local time for making the connection.

**After the flight**   The important message is to adopt the new local time for times of sleep, being awake, and taking meals.

*Sleep*   Since the new local time will be later than your body clock's time, you will tend to feel ready for sleep too early and wake too early. Recommendations are:-

- If you feel tired during the day, resist the temptation to take a nap. For the first day or so after the flight you might find it helps to retire to bed one or two hours earlier than your normal time — but no earlier than that!

- Try to sleep in surroundings that are as quiet, dark, and comfortable as possible.

- If you wake early (often because you need to pass urine), after doing so, go back to bed and stay there quietly until the correct rising time.

*Meals and drinks*   Recommendations are:

- Take meals of the 'correct' type (breakfast, lunch, and so on) by the new local time.

- Some believe that high-protein foods (fish, meat, cheese, and so on) and caffeine-containing drinks (coffee, tea) are best taken in the morning, and that a light snack rich in carbohydrates (fruit juice, pasta, and dessert) is best for supper. Caffeine-containing drinks should be avoided just before sleep (they act as stimulants), but they might be useful in the early evening to ward off fatigue.

- Alcohol is a two-edged weapon when it comes to a 'nightcap'. It sends you to sleep, but the sleep tends to be of poor quality, and it also acts as a diuretic, causing you to waken to pass urine.

*Physical activity, social activities, and exposure to bright light* Bright light, exercise and brisk walks, and being actively engaged in doing something are the most powerful tools for promoting adjustment of the body clock. They not only liven you up but they also make you ready for sleep at night.

It is, however, important to take them at the right time. As was mentioned in Chapter 8, a particularly good time to be outdoors and active in the first two to three days after the flight (or indoors but close to a window) is the period that corresponds to between 21:00 and 03:00 on 'old time', since this will help to delay your body clock. By the same token, relax indoors and away from windows (so avoiding bright light and activity) for the first two to three days after the flight during the period corresponding to 05:00–11:00 on 'old time', as this helps to advance your body clock. Translating these 'old' times into the new local time requires a bit of care, but this has been done for you in Table 11.2.

It will be noted that Table 11.2 indicates that bad times for light and activities tend to be in the night, and good times, in the daytime. In other words, 'When in Rome, do as the Romans do'. Synchronizing your life with that of your new colleagues actually helps the body clock to adjust. This fact — that acting in accord with others in the new time zone helps the process of adjustment of the body clock — is one reason why jet lag is rarely as bad after flights to the west as those to the east. Other reasons are:

- That you are asked to go to bed late, and so you can sleep in spite of what the body clock indicates;
- That the body clock tends to run slow anyway (see Chapter 2).

## Example 4: a flight to the east

After a flight to the east, the timing of your body clock is behind the new local time — and to adjust to it you will have to advance your body clock's time. Even so, the advice is similar in principle to that given for a westward flight.

**Before the flight** If possible, in the days before the flight, try to adjust at least partially to the local time of your destination, by going

**Table 11.2** Good and bad times for activites and exposure to natural light (outdoors) in the first two or three days after a time-zone transition

|  | **Bad local times** | **Good local times** |
|---|---|---|
| Number of time zones travelled to the west: | | |
| 4 hours | 01:00–07:00[a] | 17:00–23:00[b] |
| 6 hours | 23:00–05:00[a] | 15:00–21:00[b] |
| 8 hours | 21:00–03:00[a] | 13:00–19:00[b] |
| 10 hours | 19:00–01:00[a] | 11:00–17:00[b] |
| 12 hours | 17:00–23:00[a] | 09:00–15:00[b] |
| 14 hours | 15:00–21:00[a] | 07:00–13:00[b] |
| Number of time zones travelled to the east: | | |
| 4 hours | 01:00–07:00[b] | 09:00–15:00[a] |
| 6 hours | 03:00–09:00[b] | 11:00–17:00[a] |
| 8 hours | 05:00–11:00[b] | 13:00–19:00[a] |
| 10 hours | 07:00–13:00[b] | 15:00–21:00[a] |
|  | (Alternatively, treat this as 14 time zones to the west, see above[c]) | |
| 12 hours | Treat this as 12 time zones to the west, see above | |

[a]This will tend to advance your body clock
[b]This will tend to delay your body clock
[c]Choose one of the options — the one that suits you better — but it is essential that you stick to one option only

to bed one or two hours earlier each night and getting up one or two hours earlier each morning. Again, this tactic might not always be possible, but going to bed and getting up earlier are not likely to intrude too much upon your normal working life.

**During the flight**   The advice is the same as that given for westward flights — though, of course, whether people living in the new time zone are awake or sleeping will differ. Again, in-flight arrangements for meals, not always timed to coincide with mealtimes at your destination, will often thwart your good intentions!

**After the flight**   Once more, it is important that you adopt the new local time for times of sleep, being awake, and taking meals.

*Sleep*   Since the new local time will be ahead of your body clock's time, you will tend to have difficulty in waking up and feeling alert during the morning. You will also tend not to feel tired by the time the local inhabitants are going to bed.

The advice is the same as after flights to the west, except that now the 'dangers' are:

- If you cannot get to sleep at night, stay in bed and rest. If this continues to be unsuccessful, then some travellers purposely 'miss' the first sleep in their new time zone to make sure that they are tired on the second night.

- If you feel tired when it is time to get up — indeed, you might just have got to sleep! — do not stay in bed, but get up. A 'lie-in' of one or two hours, but no more, is permissible for the first day or so.

*Meals and drinks*   Recommendations are exactly as those given for westward flights (p. 146).

*Physical activity, social activities and exposure to bright light*   All the advice about activity and exposure to light given for westward flights (p. 147) applies equally after flights to the east — except that the times are different. Now you wish to advance the timing of your body clock. Therefore, activities outdoors (or sitting indoors next to a window) during the period corresponding to 05:00–11:00 on 'old time', and relaxation indoors (staying away from windows) during the period corresponding to 21:00–03:00 on 'old time', are both recommended. Again, these times have been translated into local times in Table 11.2.

Jet lag is worse after eastward than after westward flights for three reasons. First, the body clock tends to 'run slow', and so advancing it is more difficult. Second, the local inhabitants appear to go to bed too early, and so you do not feel tired at bedtime. Third, as Table 11.2 indicates, going outdoors and being active in the morning tends to make the body clock adjust in the wrong direction. This does not mean that the morning should be spent in bed, even though you might feel tired. You should get up, have breakfast, and try to stay indoors away from the windows. If you must venture outside, then use as strong a pair of sunglasses to limit your exposure to bright light.

This bias of the body clock towards delaying explains why, if the time-zone transition is of 10 hours or more, then using activities and light to **delay** the body clock is either a viable alternative (the case of a flight across 10 time zones to the east) or is to be recommended (journeys from western Europe to the antipodes). This point is included in Table 11.2.

## 11.4 How long might jet lag last?

Table 11.3 gives a guide to the duration of jet lag that might be expected after different time-zone transitions. It is based on the finding that the body clock can advance the equivalent of 1–2 time zones per day, and can delay the equivalent of 2–3 time zones per day. It also shows the large inter-individual variation that can exist. Indeed, for some individuals, the effects of jet lag might not go away completely until the number of days spent in the new time zone corresponds to the number of time zones crossed. Those who follow the above advice will hope (and can reasonably expect) that they will be towards the lower end of the ranges given (that is, adjustment will be faster).

It has been suggested that jet lag is worse in older people and in those whose lifestyle is normally very regular and ordered. In practice, there is little evidence to support these claims, and even some evidence to indicate that older travellers, and those who are experienced in intercontinental flights, have less problems, being better able to 'pace themselves'. It has also been suggested that 'larks' should be at an advantage over 'owls' when eastward flights are concerned (because they are better at advancing their body clock), and the opposite would hold for westward flights (because 'owls' are better at delaying their body clock). Again, the evidence for this is not strong.

## 11.5 Medications and jet lag

### For those on regular medication

Clearly, if you are required to take medication regularly, the instructions 'first thing in the morning' or 'with meals' will involve some

**Table 11.3** Estimates of how long jet lag will last

| Westward flights | | Eastward flights | |
|---|---|---|---|
| **Time zones crossed** | **Days to adjust** | **Time zones crossed** | **Days to adjust** |
| 0–3 | 0[a] | 0–2 | 0[a] |
| 4–6 | 1–3 | 3–5 | 1–5 |
| 7–9 | 2–5 | 6–8 | 3–7 |
| 10–12 | 3–6 | 9–11 | 4–9 |

[a]Day of rest recommended to recover from travel fatigue

irregularity after a time-zone transition and change of routine have taken place. Unfortunately, it might not be just a matter of rescheduling waking time and mealtimes along the lines suggested above, since the body does not adjust immediately to the new routine. Also, after a flight to the west, there appears to be 'extra' time, and this poses problems for the insulin-dependent diabetic, just as does the 'loss' of time and meals after an eastward flight. Consult your doctor if you are at all in doubt.

## Drugs and other aids for overcoming the difficulties of jet lag

Several possibilities come into this category. Here we will summarize the position briefly. There are the hypnotics (drugs that help you sleep) and the stimulants (drugs that wake you up). We would always advise caution in the use of these, because they might have unpleasant side-effects, and any unnecessary use of drugs is undesirable. Nevertheless, if a good sleep is imperative — say, before an important meeting — then a short-acting hypnotic might be helpful (it needs to be short-acting or its effects might last into the meeting). An alternative is to take a short nap before the meeting, making sure there is at least one hour between it and the end of the nap, so that you can waken fully. Alcohol is a poor way of getting a long sleep as its diuretic effect causes you to wake up because of a full bladder. Stimulants in the form of pills are unnecessary; coffee, fresh air, and light exercise are effective substitutes.

Pills can be bought which, it is claimed, can help adjustment of the body clock. One example of this is a treatment that requires two pills to be taken each day, one in the morning and the other in the evening. Each pill contains several substances, but among them are the amino acids tyrosine (in the 'morning' pill) and tryptophan (in the 'evening' pill). In effect, these are the 'active constituents' of a dietary regimen that could act as a zeitgeber. This diet has been discussed briefly in Chapter 8. We would repeat that the dietary regimen has not been tested satisfactorily, and we are not aware of any published scientific study relating to the use of these pills.

Last, but by no means least, there is the use of melatonin. There is now a large body of scientific evidence that shows that in many, but not all, studies, subjects taking melatonin suffer less jet lag than those living identically but taking a dummy pill (placebo) instead.

Normally, the pill is taken a few hours before sleep is desired, and it seems that melatonin is acting as a mild hypnotic, a role it appears to have in subjects living normally. (It might be recalled from Chapter 3 that melatonin is normally secreted into the bloodstream in the evening, and possibly promotes sleep at this time by accentuating the fall of body temperature produced by the body clock at this time.)

The effect of melatonin in reducing the other symptoms of jet lag has been studied far less, and the results are not conclusive. However, in a study performed on subjects who were living normally and had not undergone a time-zone transition, melatonin capsules were ingested in the early evening. There did not seem to be any residual effects of melatonin the following morning, as the subjects did not report any fall in alertness, mental performance, or physical performance at this time.

In spite of all this positive evidence, we cannot recommend the use of melatonin, unless it is under medical supervision. Why the reticence? There are several reasons:

1. Melatonin does not have a licence and it cannot be bought in the UK and many other countries. In the USA it is more freely available, being classed as a food additive.

2. The melatonin capsules that are commercially available rarely come in a pure form, but rather in combination with other substances, presumably to further improve the health of the person taking them. What effects these other substances might have, in combination with melatonin, have not been studied in detail.

3. Melatonin has not undergone long-term toxicology trials. This is hardly surprising since they would be expensive, and melatonin, being a natural substance, cannot be patented. Therefore, the safety of melatonin cannot be assured, at least as far as potential litigation is concerned. However, drugs related to melatonin are being produced, and these can be patented. Early reports with regard to their efficacy in dealing with jet lag are promising.

4. There is no medical obligation for any side-effects from taking melatonin to be reported. Occasionally, there are reports of unwanted side-effects, such as nausea and undue fatigue, but the medical authenticity of such reports is uncertain, and the frequency of such problems is unknown. Even the advocates of melatonin advise against its use by expectant mothers and young children, though this is likely to be precautionary rather than based on strong medical evidence.

For those who wish to use melatonin, we again strongly advise that they first consult a doctor.

## 11.6  Check-list for dealing with jet lag

1. Check if your journey is across sufficient time zones for jet lag to be a problem (see Table 11.1 and associated text). If it is not, then refer to advice on overcoming travel fatigue (see Example 1 (p. 144) and Chapter 10).

2. If jet lag is likely, then consider if your stay is too short for adjustment of the body clock to take place. If it is too short, then see Example 2 (p. 144).

3. If your stay is not too short and you wish to promote adjustment, then see page 145 and Example 3 (westwards time-zone transition) or page 145 and Example 4 (eastwards time-zone transition). In both of these cases, the advice relates to:

   - Before the flight
   - During the flight
   - After the flight

   The most important advice relates to after the flight.

4. The advice for promoting adjustment concentrates on:

   - Sleep
   - Meals and drinks
   - Physical activity, social activities, and exposure to bright light (Table 11.2)

5. Implementing the advice given in this chapter will not remove jet lag, but it can reduce it towards a minimum (see Table 11.3) — a minimum that will depend upon you as an individual.

Enjoy your flight and have a good trip!

# Chapter 12

## Night work — the problems

Biologically, we are essentially daytime (diurnal) creatures — awake in the daytime and asleep at night — and most of us work in the daytime, at a '9-to-5' job. About 20 per cent of the workforce suffer some biological abnormality, however, insofar as they work at night. Such an 'abnormal' pursuit affects many branches of the workforce. It is found in broadcasting, the hospital service, the fire-fighting and police forces, and in military personnel, computer operators, bankers, food distributors, long-distance lorry drivers, workers in the chemical, mining, refining, and steel industries, catering staff, hoteliers, coastguards, and many others — the list is almost endless.

Often night work is done as part of a shift system in which all shifts — morning, afternoon, and night — are worked in rotation. For others, the hours of work might be abnormal, but are regularly so. Thus, bakers and nightclub workers will all routinely work during some part of the night. There are also some 'permanent' night workers — individuals who work only at night (though they have rest days and holidays, of course). Common examples are night-watchmen and machine operators in the newspaper industry. To describe this state of affairs, it is often said that we are part of a '24-hour society'.

Even though we shall concentrate on night work and the night worker in what follows, we will refer to other shifts where they result in some particular difficulty. The main problems relate to accidents and public safety, to the quality and quantity of work done, and to the workers themselves (their social and domestic affairs, and their health).

## 12.1 Concerns of management and the general public

There are several reports that indicate that the quality and quantity of the work done on night shifts is lower than on other shifts, and that accidents are more common or serious then. There might be several reasons for this. First, for those who work outdoors, there are obvious differences with regard to the lighting and the weather conditions. Driving at night can be more difficult in the dark and when there is a greater chance of ice on the roads; but also it may be more difficult because there is less traffic and so boredom becomes a problem. For those who work indoors, there is the need for artificial rather than natural lighting; but there might also be other, more subtle problems (less supervision, for example). Related to this, it is not unknown for conveyor belts to move more slowly at night (!). Also, repair or maintenance of faulty equipment is likely to be slower at night due to less maintenance personnel being available.

Finally, but not least, in the context of this book, there are the human factors. Workers will be tired and might also be deprived of sleep to some extent. Taking a nap during night work appears to be more common than some (excluding possibly the work force!) might plan or think desirable, though there are some factories in which a nap is incorporated into the shift schedule. Whilst this break improves subsequent performance, it is essential that the effects of 'sleep inertia' (poor performance immediately after waking up) have been overcome before important processes are undertaken or decisions taken.

It is not difficult to see that this combination of factors might lead to poorer workmanship, to more errors, and even to a greater incidence of accidents. Such errors and accidents might have far-reaching consequences. Consider, for example, the potential effects of errors of navigation, of passing train signals at red, or of supervision of nuclear or chemical plants.

## 12.2 Problems of the night workers themselves

### Personal and social problems

The effects of night work upon workers permeate many aspects of their lives, both in the short term and more chronically. Night workers have to make the most of what can be unsatisfactory family and social circumstances. Obviously, they are as likely as anyone else to be married with a family or to be single, and to have friends and the need for a social life. Since family and community life are normally orientated in accord with the leisure and working hours of the majority, night workers are often excluded from sharing family mealtimes and leisure activities, or from participating regularly in sporting, cultural, or political pursuits. Even daytime television caters for a different audience compared with times of peak viewing. Sexual problems and difficulties with interpersonal relationships are more common, and there is less opportunity to share child-rearing responsibilities.

Night workers might, therefore, feel a sense of isolation. Leisure time and rest days need to be used carefully and constructively to maintain contacts with the partner, family, friends, and neighbours. There is a tendency for friends to come from a social circle of those who also work at night, since shared or common problems can draw people together.

In addition to the frequently better pay, there are some advantages to night work which should not be forgotten. These include access to shops and other services during less busy hours, the opportunity to follow solitary pursuits such as golf, gardening, and fishing (which day work seldom permits), and the possibility of more rest days, allowing time to 'get away from it all'. It is probably these considerations that lead to some night workers actually enjoying night work. Even so, it is usual for them to revert fully to a conventional, day-orientated lifestyle during rest days. Importantly, part of the advice given later (Chapter 13) will be to consider the positive rather than the negative aspects of night work.

## Health problems

There are two main areas of complaint. The first centres on gastrointestinal disorders and cardiovascular problems. The symptoms of gastrointestinal disorders include loss of appetite, indigestion, and bowel irregularities. In the longer term, gastric and duodenal ulcers are more frequently experienced by night workers than by their colleagues who work during the daytime only. These disorders are blamed (without much firm scientific foundation) on the 'abnormal' hours of work, but there are other factors that will contribute. The catering facilities at night might be absent or poor, with little choice, and diet might be dominated by 'convenience' foods. Possibly as a result of this, some night workers continually nibble snacks rather than have a proper meal. In addition, some might consume larger quantities of coffee in a bid to stave off drowsiness, and others might use alcohol to promote sleep after the night shift.

At home, the position might be little better, the individual having to eat separately from the family. The result is a decreased desire to prepare full and wholesome meals.

The truth is that, at the moment, we are unsure what combination of these factors — general 'stress', the body clock, habits, a loss of appetite, an inadequate selection of food, poor or absent eating and cooking facilities — is responsible for the gastrointestinal problems that are experienced. Finding this out (see the 'Now try this' section at the end of Chapter 6) will enable better and more appropriate advice to be given to the workforce, their partners, and management.

With regard to cardiovascular problems, there are now several reports of increased risks of heart attacks and strokes, and of an increased incidence (about one-third more) of arteriosclerosis (hardening of the arteries) among night workers, compared with day workers. This increase appears to be independent of other possible contributing factors — differences in weight, exercise, smoking habits. The reason is unknown; to cite 'stress' is an easy option, though this raises a question as to the exact cause of the stress. Alternatively, or in addition, differences in some aspect of diet might be involved. Whilst the outcome — an increase in the frequency of cardiovascular problems in night workers — is not in dispute, further work is required to establish in detail the reasons for this.

The second area of complaint centres on sleep loss, with night workers commonly complaining that they have difficulty in obtaining sufficient sleep during the daytime. On average, they get at least one hour less sleep per daytime sleep compared with their normal sleep at night. Thus, even if only a 5-day week is worked, the night worker will have accumulated a sleep deficit of 5 hours. Such a loss of sleep is not unexpected, since the time when the night worker has the opportunity to sleep tends to clash with conventional times for social and domestic responsibilities. This might be a particularly marked problem for any night-working parents with young children at school, if they have to prepare the childrens' breakfast and take them to school before going to bed. In general, normal domestic routines will tend to be disruptive for a night worker attempting to sleep, because housework is often noisy. Also, the partner and children can find it inconvenient to remain sufficiently quiet. The problem of being woken up is also exacerbated by external factors — traffic, incoming telephone calls, neighbours and their children, and so on — which cannot always be controlled.

This loss of sleep is made up to some extent during rest days, and when the worker is on daytime shifts, by the more frequent use of naps. Even this, however, can be a cause of friction, if it is assumed that days without work are an opportunity when the family and partner can all do something together.

The morning shift, starting at about 06:00, can be almost as great a cause of sleep loss as the night shift. This situation arises because workers tend to maintain some social life in the evenings, and so go to bed too late to enable them to get a full night's sleep before having to rise at between 04:00 and 05:00 the next morning. Even if sleep is attempted at a 'correct' time (about 21:00), noise and a body temperature that is still too high at this time might thwart such good intentions.

Such sleep loss can lead to chronic fatigue, which can be debilitating both with regard to performance at work and to spending leisure time effectively. Thus, ways to reduce fatigue become an important part of a night worker's 'armoury'.

## 12.3 Night work and the body clock

Many of the above difficulties arise because of a clash between external factors and the individual's lifestyle and conditions of work. There is an additional factor lurking around too — one due to the fact that, biologically, we are diurnal, not nocturnal, creatures. This internal factor is, of course, our daily rhythms and the body clock.

During night work, our body clocks will be telling us that we should sleep. We will feel tired, find work difficult and irksome, and have difficulty in concentrating (see Chapter 5). If there is a substantial physical component in the work, this will appear more tiring than usual (Chapter 4), and if there is one that requires manual dexterity then, compared with working in the daytime, we will be 'all fingers and thumbs' (Chapter 5). Our appetite and digestive system will be sending conflicting information. Although we will feel that we should eat, our appetite will be poor and we will not want a full meal (Chapter 6) — given it is possible to get one in the middle of the night. By contrast, during the daytime, even if our bedroom is comfortable and quiet, and the rest of the household accepts that we need to sleep, we will find we cannot sleep (Chapter 3), or that it is fitful. We might have to get up to empty our bladder or bowels (Chapter 6).

All these daytime problems arise because the body clock is now trying to wake us up for a new day. In other words, when we start night work, our body clock manifests its stability and continues to remain adjusted to 'normal' habits — daytime wakefulness and activity, and night-time sleep.

## 12.4 Differences between the night worker and the time-zone traveller

Problems associated with this slowness of adjustment of the body clock have already been met when we considered the problems facing the traveller across time zones. The night worker, like the time-zone traveller, needs to adjust his/her body clock to the altered sleep–wake cycle. However, it is found that, whereas adjustment to time-zone transitions is substantial after a few days (see Chapter 11), for night work, it is slower (see Fig. 12.1). Figure 12.1 shows the rhythm of excretion of adrenaline in the urine in a group of subjects during day and night work. (It will be recalled from Chapters 2 and 4 that adrenaline levels are a measure of 'activation'). When active in the daytime and sleeping at night (section a of Fig. 12.1), the subjects show a clear rhythm of adrenaline secretion, with high values during daytime work and low values during sleep at night. During the first week of night shifts (section b), adjustment of the rhythm is only partial. Thus, adrenaline is much too high during the daytime to allow unbroken sleep and, though raised during the night, it is not as high during these new hours of work as it was when work was being done in the daytime. By the third week of night shifts (section c), adjustment is better, insofar as the values during daytime sleep are low. However, values during work at night are still not as high as those during daytime work (see section a).

Adjustment is slower because there is a clash between different zeitgebers. Individuals might adjust habits to fit in with the demands of night work, but they are aware that, in doing so, they are at variance with society as a whole. Moreover, the natural light–dark cycle will be in opposition to the night worker's artificial light–dark cycle.

The problem is worse than this, however, for several reasons.

1. During rest days, night workers generally revert to a conventional lifestyle (often due to all the surrounding social constraints); now all zeitgebers adjust them to this. This means that any hard-won adjustment of the body clock to night work tends to be lost quickly during rest days or days when the day shift is worked. This is shown in section d of Fig. 12.1 — the rhythm of adrenaline in the first week back on day work becoming almost identical to the original rhythm (section a). Unless the individuals are social 'loners' or part of an isolated community where night work is the norm, they

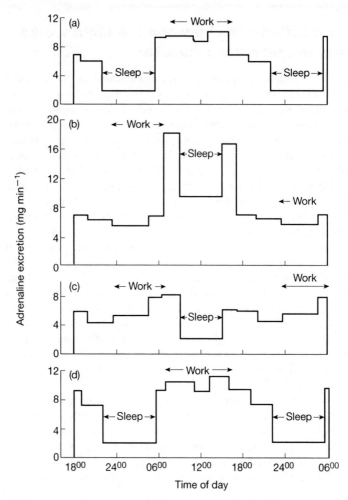

**Figure 12.1** The daily rhythm of adrenaline excretion in the urine: (a) during day shifts; (b) during the first week on night shifts; (c) during the third week on night shifts; (d) during the first week back on day shifts.

will always be subjected to competing influences with regard to the timing of their lifestyle and the body clock.

2. Even if adjustment of the body clock to night work were to take place (with performance rhythms — both mental and physical

performance — and the body temperature rhythm peaking at night, and showing minima during daytime sleep), there would be another problem. On day work, we work during the times when body temperature and performance are rising and reaching their peak, and we relax after work during the times that these rhythms are falling and getting us ready for sleep. On night work, by contrast, we tend to take leisure first (in the afternoon and evening, so we can see our family and friends) and then work afterwards, in the second half of our waking time (Fig. 2.4). This means that we are working not only when the daily rhythms of performance and body temperature are near their troughs, but also when fatigue levels are higher because of the greater length of time that has passed since the last sleep (see Fig. 3.1, bottom, arrow B). This effect of increased fatigue could be overcome if the night worker adopted the sequence we adopt during day work — namely, sleep–work–leisure. This would entail sleeping in the late afternoon and evening, and using the morning for leisure purposes. This pattern would, however, put the night worker at a severe disadvantage with regard to being with their partner, family, and friends. There is, thus, a clash between individuals' biology and what is socially desirable to them.

3. The problems due to night work can continue throughout the individual's working life. Every time that night work begins, another change of routine and a temporary mismatching between lifestyle and the body clock will occur. The continual changes of work pattern will mean that the body clock can never settle down. An analogy would be a jet-setter forever crossing from one time zone to another and moving on before adjustment to the current time zone had taken place. It might seem that aircrew involved in long-haul flights across multiple time zones would have similar experiences, but, in practice, this is not often the case. This is because there is a tendency for older pilots and other members of aircrews to choose to opt out of such flights, choosing, instead, journeys that involve a smaller number of time-zone transitions.

## 12.5 Differences between individuals

From the above, it is not surprising that some workers leave night work! Those who suffer most from its undesirable side-effects tend to transfer to day work rather than to endure long spells of suffering from 'night workers' malaise'. In general, it becomes less usual for a night worker to be more than 50 years of age. The fact that those who suffer more than average from any ill-effects are more likely to leave night work has an interesting implication when attempts are made to quantify any undesirable effect of night work. It means that, as we look at older or more experienced night workers, we are dealing progressively more with a self-selected group — those who, by luck or by the development of 'coping strategies', are unrepresentatively healthy. (There is an analogy here with the idea of 'survivors', when old age is considered. See Chapter 7.) Looking at only those who continue night work might lead us to underestimate the size of any problem. To get a better estimate of the true size of the problem, we need to include those who have left night work. Some would argue that it is **those who have left night work** who are the most valuable indicators of what can go wrong.

In spite of all the possible difficulties, most workers do not leave night work, at least not while they are under 40 years of age. For some, the problems of night work appear to be minimal, these being offset by the advantages it can offer in some cases (more free days, more money, greater autonomy, and, possibly, more rapid promotion).

If, as indicated above, individuals differ in their ability (or willingness) to work at night, can those who will have least or most difficulties be predicted? If so, the former group might act as a model for advising people in general about night work; and the latter can be warned against it, or given detailed advice on how best to deal with it. It must always be borne in mind, however, that, even though it might be possible to change some differences, there might be others (age, for example) about which nothing can be done.

The main differences between individuals are explained below, though there is little supporting evidence. It is better to consider them as possibilities that are based on our current understanding of the problems associated with night work, and on what are believed to be their causes.

## Age and experience

With age, some people become less flexible (Chapter 7), and this is an argument against undergoing the continual changes of routine that night work entails. The evidence above, about a decreasing tendency to work at night with increasing age, supports this view. However, older people are often better at 'pacing themselves', and this might be as valuable in night work as it appears to be after a time-zone transition (see Chapter 11). Older workers will also have more experience of dealing with night work, and might be able to give advice to younger members of the workforce.

## Motivation and fitness

Those who are motivated to do night work might have learned 'coping strategies'. There are some reports of positive effects upon the workforce of having taken part in training sessions designed to increase physical fitness. Whether the benefits stem from the increase in fitness itself, from the associated increase in social life, or from a more general placebo effect is unclear. To some extent, it does not matter; if it works, who cares what the mechanism is?

## Flexibility of sleeping and waking habits

People differ in their ability to take a nap when the opportunity arises, and in their ability to sleep at the 'wrong' time (daytime, in this case). Also, they differ in the ability to overcome feelings of drowsiness, and in the ease with which they can turn over and go back to sleep (rather than get up and complain) when woken during a daytime sleep. These differences reflect the psychological make-up of an individual, and so might not be easy to alter. Nevertheless, it is possible that those who are flexible about these issues can pass on information about how they manage to deal with them.

## The timing of daily rhythms

The differences between the habits of 'larks' and 'owls' have been dealt with in Chapter 1. Since all daily rhythms seem to be affected similarly, it would be predicted that larks would be better suited to a morning shift, not only because they are better able to work in the morning, but also because they would find it easier to go to bed and get to sleep early in the evening, so enabling them to get sufficient sleep. By contrast, the owls would appear to be better suited to night work, since they can stay alert and work later into the night, and can get more sleep in the daytime after working the night shift.

In practice, with rotating shift systems (in which all shifts are worked in sequence by the workforce), neither group is advantaged. The obvious solution — to allow the larks to do most of the morning shifts, and the owls to do most of the night shifts — does not appear to have been tested, though the increasing use of 'flexitime' could accommodate this possibility. When only one shift is worked, and the individual has a choice as to which one it is, it seems likely that the selection might be determined, in some individuals, by the extent to which they are a lark or owl. Choosing to work morning shifts if one is an owl, or night shifts if one is a lark, is unlikely, and would seem inappropriate (see Chapter 1).

Armed with this information about the problems of night work and differences between members of the workforce, we must now try to give advice to all concerned. This is in the next chapter.

# 12.6 Check-list for the difficulties and advantages associated with night work, and differences between individuals

**Difficulties for management and the general public**

- Poorer quality and quantity of work (when this can be measured)
- An increased frequency of errors and accidents

**Difficulties for the workforce**

- Decreased social, family, and personal life
- Chronic fatigue and indigestion
- Increased incidence of ulcers and cardiovascular disorders

**Advantages to management and society**

- More effective use of machinery
- Availability of services around the clock

**Advantages to the workforce**

- More money and autonomy
- Spare time when it is less busy

**Differences between individuals**

- Age and experience
- Motivation and fitness
- Flexibility of waking and sleeping habits
- The extent of being a 'lark' or an 'owl'

# Chapter 13
## Night work — advice

In Chapter 12, we considered, and explained where possible, the problems that are associated with night work. We have also discussed those factors that might affect the ease with which individuals can adjust to it. On the basis of this evidence, we can now offer advice with respect to the least troublesome types of shift schedule involving night work, and the means by which a worker can deal best with a particular schedule. This advice is for the workers, their partners, families, and friends, as well as their colleagues at work. It also concerns the kind of work undertaken at night, and the facilities provided at that time.

First, we must consider if there are medical grounds that might cause a worker to take professional advice before working at night.

## 13.1 Groups who might decide against night work on medical grounds

There are certain groups who would be advised to think very carefully before performing night work and to seek medical counselling. These groups include the following:

### Epileptics

Epileptics are more susceptible to seizures when fatigued and so cumulative sleep loss should be avoided. The morning shift (from about 06:00 to about 14:00) is also likely to be associated with sleep loss, particularly with 'owls', who will have difficulty in getting to sleep early enough the previous evening.

### Asthmatics and those with respiratory disorders

For anyone, allergic reactions (to house dust and the like) are often worse overnight (see Chapter 14). This is due in part to a low blood concentration of the hormone, cortisol, during the late evening and early night (see Fig. 3.4). Asthma is no exception to this rule, and substances at work, such as dust, lint, and chemicals might worsen the position.

*Insulin-dependent diabetics and others taking regular medication*

As was the case following a time-zone transition (Chapter 10), there is a problem in interpreting an instruction such as 'three times per day with meals' or 'once a day on rising' if the schedule is continually being changed. The insulin regimen of diabetics will be very difficult to judge accurately if there are irregular mealtimes, and so they would be advised against shift work in general, and night work in particular, for this reason.

As a further example, arthritic pain is often worst in the morning for those working and sleeping at conventional times, again, partly due to the effects of a lack of cortisol during the night (Chapters 3 and 14). Some medication is often taken before retiring, to reduce the pain and swelling on waking the next morning. With a changed sleep–wake schedule, therefore, should the individual be advised to take the medication always before bedtime or to take it at a time coincident with the time of lowest concentration of cortisol in the blood? If the latter advice is given, how will this time be found on a particular day?

**We reiterate that individuals are encouraged to seek medical advice if any of the above, or related, issues cause them concern.**

## 13.2 Advice on the organization of shift systems

Night work is generally part of a shift system, and changing one shift will have a 'knock-on' effect for the others. Therefore, we have to consider shift systems as a whole. Some of the advantages and disadvantages of different shift systems are summarized in Table 13.1.

There are many types of shift system, often designed for a specific workforce. Nevertheless, from our viewpoint, we can consider them with regard to differences in the number of hours worked each shift, the times when the shifts start, the direction of rotation of shifts, and their speed of rotation. The comments below are based on chronobiological principles. Unfortunately, there is no clear epidemiological evidence to indicate whether one type of system should be preferred to another.

**Table 13.1** Some advantages and disadvantages of different shift systems

|  | Advantage | Disadvantage |
| --- | --- | --- |
| **Length** | | |
| 8 hours | Less deterioration | Going to work more often |
| 12 hours | More deterioration | Going to work less often |
| **Start time** | | |
| 05:00, 13:00, 21:00 | Early end to shifts | Early start for morning shift |
| 07:00, 15:00, 23:00 | Late start for morning shift | Late end to shifts |
| **N.B. Can 'flexitime' be instituted, to accord with individual preferences?** | | |
| **Direction of rotation** | | |
| Morning/afternoon/night | More time when shifts change | Early start to new cycle |
| Afternoon/morning/night | More rest time between cycles | Little time when shifts change |
| **Speed of rotation** | | |
| Slow (>2 weeks on each shift) | Can adjust | Possible sleep loss |
| Weekly (1 week on each shift) | Socially convenient | Sleep loss and cannot adjust |
| Rapid (1–2 days on each shift) | Little sleep loss | No adjustment |

## How long should the shift last?

Conventionally, shifts last 8 hours, but there has been a trend recently towards 12-hour shifts, normally divided between 'night' and 'day' shifts. A 12-hour night shift raises the question, 'What are the effects of extended work hours?'. The following comments summarize the position that was described in more detail in Chapters 3 and 5.

- For complex or repetitive, boring tasks, particularly those requiring vigilance, performance is more likely to deteriorate, or at least be harder for the individual to sustain, with an extension of work time from 8 to 12 hours. For 'interesting' and varied tasks, by contrast, performance is likely to deteriorate less during these extra hours.

- In shifts of any length, performance, especially in complex or boring tasks, or in those requiring vigilance, tends to worsen if an individual is suffering from sleep loss, or when the daily rhythms of body temperature and adrenaline secretion are near their trough, as would be the case at around 03:00–05:00.

The implication of all this is that, particularly with 12-hour shifts or if sleep has been lost, attempts to make the task as interesting as possible, and to change the type of task being done during the course of the shift, should be rewarding to the workforce and to management alike.

## When should the shifts start?

A common arrangement for a three-shift system is for shifts to start at 06:00, 14:00, and 22:00. One difficulty with this arrangement is the earliness of the start of the morning shift, which means that workers will often have to get up at between 04:00 and 05:00 in order to reach work on time. This is more acceptable for 'larks' (morning types) — since they will get to sleep early enough the previous evening — but it is rather early for most of us, and particularly so for 'owls' (evening types). As has been mentioned already, sleep loss and fatigue are often associated with the morning shift as much as with the night one.

Starting the morning shift at 07:00 (or even later) will alleviate this problem, but some workers dislike the loss of leisure time in the afternoon that this later start would entail. Also, there will be a 'knock-on' effect with the afternoon shift (15:00–23:00), severely limiting the opportunity for these workers to meet friends after work. Also, the later finish of the night shift (23:00–07:00) will mean that, by the time the workers have got home and had a meal, their body temperature and adrenaline rhythms will be well into the rising phases, so making it more difficult for them to get to sleep.

A more 'adventurous' scheme, making use of 'flexitime', would involve an arrangement whereby individuals could choose to work one set of shifts (starting at 06:00, 14:00, etc.) or another set (say, 07:00, 15:00, etc.), depending upon personal preference. This would better accommodate domestic arrangements and the degree to which an individual was a 'lark' or an 'owl', and, with cooperation between all concerned, continuous working could still be maintained.

## In which direction should shifts rotate, and when should rest days be taken?

The answer to the first part of this question depends upon whether adjustment of the body clock to the shift cycle is to be attempted or not (see below). If adjustment is to be attempted, then rotation should be in a delayed rather than an advanced direction. That is, the sequence of shifts 'morning–afternoon–night' is preferable to that of 'afternoon–morning–night'. The reason for this choice is that the body clock, with its inherent period of greater than 24 hours, will adjust more quickly to delaying than advancing shifts (for the same reason that westward flights are associated with less jet lag than eastward ones, see Chapter 11). By contrast, if adjustment is not going to be attempted, then the direction of rotation is unimportant with respect to the body clock.

There is yet another factor that needs to be borne in mind — one that applies whatever the speed of rotation of the shifts. A workforce often prefers the shift system to rotate in the sequence afternoon–morning–night, since there is a longer break between the end of the night shifts and the first (afternoon) shift of the next cycle. However, such a direction of rotation also means that the switches from afternoon to morning shifts, and from morning to night shifts, are accompanied by a break between shifts of only 8 hours. Particularly if travel to and from work takes up a substantial amount of time, this regimen allows insufficient time for meals and sleep between these shift changes. The direction of rotation morning–afternoon–night is less popular, because the first shift of each cycle is the morning shift, and this requires an early start after rest days. However, this system has the advantage that the switches from one shift to another (morning to afternoon, and afternoon to night) allow sufficient time for travel, sleep, and meals. Biologically, but not socially, the recommendation is that the sequence morning–afternoon–night is preferable in both rapidly and slowly rotating shift systems.

Whichever direction of rotation is implemented, rest days should be taken immediately after the night shifts. In this way, any sleep loss that has accumulated due to poor daytime sleep during the night shifts can be made up.

## How rapidly should the shifts rotate?

When the speed of rotation is considered, the problems of cumulative sleep loss and social disruption are important, as is whether or not adjustment of the body clock is to be attempted (see below).

Permanent night work and a slow rotation of shifts (that is, each shift is worked for at least two weeks) have the potential advantage that they give the greatest opportunity for adjustment of lifestyle and the body clock to night work. A worker will benefit greatly from a partner, family, friends, and colleagues who understand and accept the difficulties involved and cooperate as much as possible. However, if adjustment is poor (because the worker cannot adjust, or chooses not to try to do so), sleep loss and social disruption are likely to be most marked.

For many people, long stretches of night work are undesirable, but weekly stretches are more acceptable. The unit of social planning is often the week, with special importance being attached to the regular occurrence of weekdays, days for shopping, etc. It is for these reasons that many prefer a weekly rotation of shifts. However, where a weekly rotation of shifts is practised, there is not enough time for body rhythms to adjust much to night work (see Fig. 12.1, part b).

In greatest contrast to the slow rotation of shifts are the 'continental' or 'metropolitan' systems, by which shifts rotate every few days, even every single day. In such circumstances, adjustment of the body rhythms is impossible and so, on every occasion that night work is being performed, the rhythms will be timed appropriately for sleep at night. This is likely to reduce performance on the night shift, especially if tasks are repetitive or require vigilance, and to make daytime sleep more difficult to initiate and sustain. However, with this rate of rotation, cumulative sleep loss and social disruption will be minimal — 'normal' work hours will be guaranteed on at least some days each week.

## The body clock — to adjust or not to adjust?

From the above, one of the issues that is relevant to the design of shift systems is whether or not the individuals involved can, or will attempt to, adjust their body clocks to night work. As has been described in Chapter 12 (and see Fig. 12.1), this is not easy. Nevertheless, successful attempts to do so have been made. Adjustment requires the working environment on the first night shift to be as brightly lit as possible from the beginning of the shift until about 04:00, and then for the lighting to be dimmed for the last hours of this shift. On the next night shift, the bright light can be extended until 05:00 and, thereafter, can last throughout the night shift.

Such a lighting regimen will delay the body clock (see Chapters 8 and 11), but there are certain provisos with this method. First, the bright light should not hinder the work being done (as might be the case if VDU screens were being watched, for example). Second, it is important that the individual does not receive exposure to bright light on the way home after the night shift. This is not difficult to achieve in the winter months in temperate latitudes but, in the summer, dark sunglasses would need to be worn. If the workforce or management were to regard these requirements as impracticable or too restrictive (and we accept this objection), then this is a further example of the clash that exists between biological requirements and what is socially acceptable to those involved when night work is considered.

If attempts to adjust to night work in this way are to be considered, then, as a general rule, it is only worthwhile if the shift system rotates in the morning–afternoon–night direction, and if the rate of shift rotation is slow, and if there is a reasonable chance that daytime sleep can be taken satisfactorily. If these conditions are not all met, then a more rapid rate of rotation is recommended, and it is must be accepted that the body clock will show little or no adjustment to the night shifts.

## 13.3 Advice to the workforce (and its implications for others)

There are several areas where advice can be given to those who work at night. They are summarized in Table 13.2.

The first pieces of advice apply whatever the shift system.

### How to combat fatigue during night work

Such a feeling of fatigue — particularly during night shifts worked as part of a rapidly rotating shift system, or during the first few nights of a longer sequence of night shifts — is to be expected. Where possible, try to rouse yourself with a breath of fresh air or a brief burst of exercise. Other methods include deep breathing, sucking a slice of lemon, and pinching oneself. A cup of coffee can work wonders for increasing arousal, but do not overdo it (see below)! Also, if possible, try to change the kind of task you are doing (can you exchange temporarily with a colleague?) so that things become less boring. During breaks, particularly the 'lunch break' in the middle of the shift, is there a quiet place to sit down for your meal and to chat to your colleagues? Can you find something to do, rather than just 'hang around'?

### Sleep and naps

We hope the need for as much sleep as possible — and this requires the cooperation of others living with you — has been stressed sufficiently. A quiet, darkened bedroom is best and, with it, the opportunity to escape

**Table 13.2** Summary of advice to night workers

---

**Combat fatigue during night work**
- Do breathing exercises
- Try a burst of exercise
- Make the most of meal breaks
- Can you alternate tasks with a colleague?

**Sleep well and make use of naps**
- Ensure the bedroom is conducive to sleep
- Try relaxation exercises rather than alcohol to get you to sleep
- Minimize the chance of interruptions (telephone, visitors, household)
- Use naps to 'top up' lost sleep

**Eat proper meals**
- Do not 'nibble' on night work — eat a proper meal (microwave?)
- Cut down on fatty foods — eat fresh fruit and vegetables instead
- Do not drink too much coffee
- Enjoy the social aspects of a meal break
- At home, try to meet the family over the evening meal

**Use of leisure time and rest days**
- Take advantage of less crowds when shopping in the daytime
- Pursue your hobby — or go out for trips when it is less crowded
- Be sure to rest and catch up on lost sleep

**To adjust or not to adjust?**
- If the shift does not change for at least a week, then try to adjust
- If the shift changes every few days, then do not try to adjust

**Learn from experienced night workers**
- How have they learned to cope?
- What 'tricks of the trade' can they pass on to you?

---

from the telephone and other unwanted distractions. Ear plugs and relaxation exercises might both be useful. Alcohol is a bad 'nightcap' before retiring, as it will tend to cause your bladder to fill and so waken you.

Be prepared to make use of naps to 'top up' lost sleep. Often, a nap at about 14:00 for an hour or so is a good idea, since most people feel tired then anyway (the 'post-lunch dip'). A nap of similar length, but later in the day just before the night shift, can also be valuable, preparing you for work (see Fig. 2.4).

## Meals

Do not skimp on meals, either at work or at home. You are advised against 'nibbling' your way through the night shift, or making do with endless cups of coffee or tea — often, also, with too many cigarettes. Too much caffeine might exacerbate any problems with indigestion. Try to structure your eating habits as much as you would during a normal daytime routine; 'elevenses', a 'lunch break', and a later break for tea should all be fitted into the night shift and become as much a part of its routine as is the case with a daytime shift. If possible, join your colleagues for these breaks, particularly 'lunch', and make sure there is time for a chat also (see above). For 'lunch', eat a fairly substantial meal. If you have access to a cafeteria, then make use of it — but cut down on chips and fried food, and concentrate rather upon pasta or salads, for example. If you do not have a cafeteria in your workplace, or it is not open at night, then a microwave oven means that appetizing and warm meals can be prepared quickly and with ease. If you have neither, then it is our view that management should be doing more to help!

At home, the rest of the household will consider that you will want the 'wrong' type of meal at a particular time. For example, you are less likely than the other members of the household to want a breakfast at 08:00, just after you have finished the night shift. Particularly if there are children having breakfast before going to school, you might have to prepare your own meal, or settle for something that is not exactly what you want. Equally, you might not want a large meal in the evening (possibly, quite soon after you have got up) before you go to work. These are problems that we comment upon at the end of the chapter.

The idea that the type of meal can help adjust the body clock (see Chapters 8 and 11) has not been tested in connection with shift work.

## Leisure time and rest days

Use your leisure time constructively, and make use of the advantages that having leisure in the daytime can offer. You can arrange to do the shopping when there are fewer crowds and less queuing, and there will be a greater latitude when it comes to arranging appointments during the daytime. Make use of it also for hobbies; 'off-peak' membership of some leisure centres is cheaper, for example.

Rest days are designed also to give you a 'rest'. They are for catching up on lost sleep, not only for catching up on household chores such as shopping, washing, and gardening! Try to use them for excursions and other leisure activities — again, away from the crowds.

The secret in all of this is to 'think positive'. Devise ways of accentuating the advantages of your predicament, and playing down the disadvantages.

## Advice dependent upon the type of shift system

With a slow rotation of shifts, make every attempt to adjust to night work. This involves your sleeping habits as well as your eating habits at work and, if at all possible, the timing of your exposure to bright light and avoidance of it (see above). Such an adjustment will require cooperation from management as well as your partner, family, and friends.

With a rapid rotation of shifts, the problem will be mainly one of overcoming the sensation of fatigue on the night shift, as you will not have sufficient time to adjust to it. To guard against the possibility of becoming disorientated — 'what day is it?' 'what should I be doing today?' — you are advised to structure your lifestyle with a conventional, day-orientated routine as far as possible. Make use of naps, but time them wisely. Do not take a nap shortly before you plan to go to bed, and make sure that, after any nap, you leave enough time to wake up properly before doing something that requires your attention (driving, for example).

If you have a few rest days in the middle of a whole series of night shifts (doing permanent night work, for instance), try to arrange for some of your sleep still to be **at the 'normal' time associated with your night work** — that is, in the daytime. This should reduce the loss of adjustment to night work that rest days often produce.

# 13.4 Lessons from the experienced night worker

Shift workers who have worked night shifts for a number of years and, therefore, who presumably tolerate it quite well, often structure their lives around the demands of night work and place social life second. By contrast, for those who find night work difficult, social life often takes precedence over sleep and mealtimes. It is not only this priority between work and social life, but also the degree of structuring of activities, that can differ however. Thus, the worker tolerant of night work may structure the day more than does one who has difficulties with it. This structuring might include regular times for shopping, housework, walking the dog, visiting the 'local', or taking meals. Without such a routine, some individuals can become disorientated. Structuring the day in this way has been described as the development of 'coping mechanisms'. It might be that they act as a set of 'personal time cues' or zeitgebers, and they might also reflect the fact that the individual is unprepared to let the abnormality of the routine required by night work disrupt a full and interesting lifestyle.

Unfortunately, such intentions can be thwarted. There is the problem of sleep loss for those living in a busy area. It is also likely that parents working on a night shift as well as attempting to care for young children during the day will lose out on sleep — both they and, ultimately, their families might suffer. A problem might exist for some workers if their culture conflicts with some aspects of the demands of night work. For example, Moslems, who are required at times to fast during the day, might have to rely on canteen facilities on the night shift to provide their chief source of nutrition.

Many of the points we have raised can be assimilated into a model that describes a 'committed' or 'motivated' night worker. He or she will accept the changes in lifestyle that are involved and attempt to make use of the advantages night work offers, rather than be irked by the disadvantages. This will require a dedication to work and a careful and positive use of leisure time. This positive attitude will manifest itself as a regular lifestyle with regard to times for sleep, meals, and chores such as shopping and appointments.

## 13.5 Conclusion and summary

The advice on shift systems and to shift workers is summarized in Tables 13.1 and 13.2, respectively. Biologically, shift work in general, and night work in particular, is unsatisfactory. Clearly, for the worker, factors such as financial reward and promotion can outweigh biological and social difficulties, and the advice given above and summarized in Table 13.2 might further ameliorate some of them. Even so, we are aware that, on occasions, we have done little more than to state, 'If you do night work, then you cannot do this' or 'If you do this, then it will make night work that much more difficult'. This is the nature of night work — a clash between the requirements of work, social factors, and biological factors.

To some extent, the ability to deal successfully with night work is up to the individual. Certain factors — including biological factors, factors resulting from the attitude of the partner, family, friends, or work colleagues, and factors arising from the work environment itself — will make things easier or harder for that individual. Ultimately, each individual must develop a positive attitude to night work, by stressing its advantages and by playing down its disadvantages.

For some, the disadvantages continue to outweigh the advantages, in spite of having attempted to develop coping mechanisms. For them, there seems no alternative but to end night work. There is the consolation that those who do so for these reasons appear often to recover. By contrast, those with the same problems but who continue to work at night will almost certainly find that their symptoms and problems worsen.

If it is accepted that night work must be carried out (and this is not seriously disputed), then it must also be accepted that some will suffer as a result. When there is evidence that an individual is having, or is predicted to have, particular difficulties with night work, this individual should be advised accordingly. In our view, all such individuals should have free access to confidential medical advice and be able to transfer out of night work (should their condition warrant this) without prejudice to their employment opportunities.

# Chapter 14

# Daily rhythms and medicine

Hospitals are day-orientated with regard to administrative and consultative business, non-emergency (elective) operations, and treatment. This is to be expected, and is for the convenience of both the staff and patients. Even so, the hospital cannot shut at night; it must continue to look after the in-patients, it must be able to provide them with medical care whenever required, and it must always be able to deal with emergencies.

## 14.1 Emergencies and circumstances that must be dealt with immediately

Emergencies might occur at any time. Nevertheless, accidents occurring in the workplace will be concentrated during the daytime (though night shifts produce more than their fair share of accidents, see Chapter 13), while accidents in the house, such as falling down stairs, might be more frequent early in the morning or late at night. Accidents due to sport will occur mainly at the weekend, on Saturday afternoons and Sunday mornings in particular, and those resulting from brawls will be mainly on Friday and Saturday evenings.

In addition, there is a daily variation in the onset and severity of symptoms in some diseases.

### Cardiovascular problems

Acute cardiovascular problems — a heart attack, blood clot, or stroke — all tend to be slightly more frequent between 06:00 and noon than in any other 6-hour period. This is believed to result from a combination of several circumstances:

- The tendency for the blood to clot is higher during this 6-hour period than at other times of the day.

- Blood pressure is rising most rapidly, from its nocturnal minimum to a maximum at about 11:00 (see the 'Now try this' section at the end of this chapter).

- Individuals are facing the mental and physical rigours of a new day.

That is, the physical demands made upon the heart are increased in the hours immediately after waking, as are the physical stresses placed upon the arteries. It is these vessels that must withstand the higher blood pressure. At the same time as the blood vessels have to deal with these challenges from an individual's environment and lifestyle, the tendency for blood to form a clot is highest. This combination of factors may well tilt the balance towards cardiovascular problems at this time of the day.

## Allergic responses and asthma attacks

The frequency of asthma attacks is highest in the evening and during the night (Fig. 14.1). The same applies to those attacks of breathlessness that cannot be attributed to physical activity.

In part, this is because the size of the airways shows a daily rhythm (see Chapter 4), being greatest in the daytime, when we need to breathe most in order to get sufficient oxygen into the lungs, and narrowest at night. High daytime adrenaline levels help to expand the airways, and the falling levels of this hormone in the evening will cause the airways to begin to constrict.

There are two further reasons for this nocturnal decrease in airway size. The first is an increase in activity of the parasympathetic nervous system. This part of our nervous system is not under voluntary control, and it acts in the opposite way to the sympathetic nervous system. Whereas the sympathetic nervous system rouses us to daytime activity, the parasympathetic nervous system quietens us down in preparation for sleep. The switch from sympathetic to parasympathetic nervous activity is one of the reasons that the heart rate is lower at night, and it also results in a narrowing of the airways.

The second extra reason is that the airways are affected by our response to allergens — substances causing an allergic reaction (dust and pollen, for example). There is a daily rhythm in allergic responses in general. Injections (under the skin or up the nose) of material causing allergic responses lead to a much more marked response (a wheal at the site of injection, itching and watering of the eyes and nose) in the evening than in the morning. The low blood concentrations of the

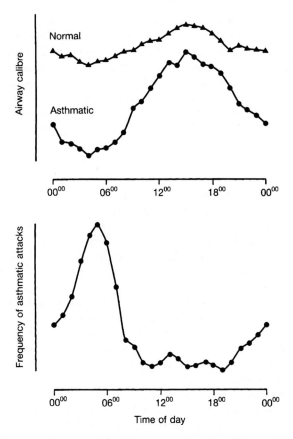

**Figure 14.1** Top: relationship between time of day and airway calibre in normal subjects and asthmatic patients. Higher airway calibre shown upwards. Bottom: relationship between time of day and frequency of asthmatic attacks. Higher frequency shown upwards.

hormone, cortisol, that are found in the evening (see Fig. 3.4) are involved in this reaction, since this hormone is known to damp down these responses. Clearly, the amount of allergen that is around is most important, as anyone suffering from hay fever will know! Thus, an allergic response can be produced whenever the appropriate substance is present, but the response in the evening is either more marked or needs less irritant, in comparison with the morning.

Whereas the amount of circulating pollen in the evening is likely to be lower than in the daytime, dust mites in our bedding (and this is not a sign of poor standards of hygiene, but rather of the ubiquity of the dust mite!) might precipitate an asthma attack in those who are susceptible. It is because such attacks — some of which require hospitalization or are even fatal — are more likely at night that asthmatics are warned against night work (see Chapter 13), particularly if an irritant such as lint is to be found in the work environment.

### Childbirth

To strike a happier note, the frequency of childbirth is higher during the later hours of the night than at other times. There are at least two reasons why the mother tends to start labour in the late evening and they show, once again, that there are both external and internal causes to a daily rhythm. The external cause is because the mother is more likely to be lying down in the evening than at other times of the 24 hours. This position promotes kicking movements of her unborn child because the pressure inside the mother's abdomen, which acts upon the foetus, changes when she alters her posture, and because the foetus is responding to a modification of blood flow from the placenta. The internal cause is that there are several maternal hormones, all showing daily rhythms, that interact to modify the strength of contraction of the uterine muscle, as well as its sensitivity to the movements of the foetus.

It will be noted that childbirth will tend to occur more frequently at night, when hospital staff are likely to be decreased in numbers and tired. In spite of this, it is reassuring to know that childbirth in the night does not appear to be associated with more problems — indeed, there is a report that births at night are less prone to difficulties than those in the daytime; the 'natural' time is best, as it were.

Not surprisingly, Caesarean sections and induced births nearly always take place in the daytime, which is no more than a reflection of the desire to plan work to correspond to the daytime habits of most of the hospital staff.

## 14.2 The time when symptoms are most marked

In many cases, patients visit their doctors because of the development of symptoms of an illness, and these symptoms might show a daily rhythm in their severity. The example of an increased frequency of breathlessness in asthmatics during the evening has already been mentioned. Another common example is the severity of a fever as assessed by the first-line approach, namely, feeling the forehead. Fevers are generally more marked in the evening, because this is when heat is being lost through the skin most quickly, as the body is cooling down in preparation for sleep (see Chapter 3 and the 'Now try this' section at the end of Chapter 4). A feverish brow is less common in the morning, when the body is tending to minimize heat loss; therefore, a fever at this time can be a clearer sign that something is wrong.

Two further examples are rheumatoid arthritis and pain.

## Rheumatoid arthritis

This is a disease of the joints in which there is inflammation, pain, and stiffness. These symptoms are worst in the morning (Fig. 14.2).

One cause of the daily rhythm of symptoms is immobility of the limb during nocturnal sleep. Therefore, staying awake and active all night might reduce the stiffness and pain. Apart from the fact that this would be neither a popular nor a sensible treatment, it would not entirely prevent the morning increase in symptoms. This is because rheumatoid arthritis is an autoimmune disease — one that occurs because the body's own immune system attacks the joints. Again, the

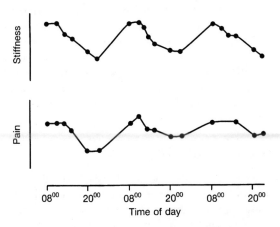

**Figure 14.2** Relationships between time of day and subjective ratings of stiffness and pain over 3 successive days in a patient with rheumatoid arthritis. Worse symptoms shown upwards.

natural immunosuppressant, cortisol, is involved. The low levels of this hormone in the evening lead to an increase in immune activity that, after a delay of a few hours, results in increased activity in some white blood cells being directed against the joints. (The response of the airways in asthma to lower cortisol levels (see above) is quicker than that in the joints; a longer sequence of events is involved in the joints.)

## Pain

Pain is probably the symptom that is most widely used by us as a sign that all is not well. In general, the threshold for acute pain (for example, the pain after a pinprick) declines as the day progresses, to reach a lowest value (that is, we are most likely to feel pain) in the late afternoon. With other types of pain (for example, an aching or dull pain from the gut), the daily rhythm is timed almost the opposite to this, and most pain is felt during the night.

In practice, thresholds and rhythms are unlikely to be a consideration in normal circumstances. When we hurt ourselves, minor differences in the amount of pain are irrelevant. For this reason also, there does not seem any point in insisting on a dental appointment at a particular time of day! However, knowledge of the time of peak pain might be relevant in the case of chronic pain, enabling more painkilling drugs to be given at the appropriate time.

Why different types of pain show different daily rhythms is unknown, though they all probably have internal and external causes. Internally, daily rhythms of naturally occurring, morphine-like substances, might play a role in determining how much pain we feel. By contrast, when we lie still (see above), when we exercise a particular joint (osteoarthritis, which seems to worsen with the amount of use of the joint), and when we eat a meal and get indigestion, are all common examples of ways in which rhythmicity in pain might arise from an external cause.

## 14.3 The time of diagnosis

In diagnosing an illness, the doctor will measure or discuss with the patient some variable that is known to become abnormal in the disorder being investigated. Having taken into account features such as the age, sex, and build of the patient, his or her recent life history, and statistical variation, the doctor can then decide if the variable is abnormal, that is, outside a range compatible with health.

Consider the example of body weight. How would we decide if a male were 'obese'? In principle, we would need to know whether his body weight was within the 'normal' range for a male of that age and height. ('Normal' is defined statistically, on the basis of measurements made upon many males, of different ages and heights. It is generally expressed as a reference range that encompasses 95 per cent of the population.) We would also have to take into account genetic and cultural effects; consider the (genetic) differences between Eskimos and Negroes with regard to build, or between inhabitants of different countries when it comes to their diet (often cultural), and the effect this has upon body weight. The past history of our male would be important also; we would expect him to be heavy if he had been undergoing athletic training, for example (due to muscular development).

Even taking into account all these factors, there would still be a range of values associated with non-obesity, just as there are ranges in hair colour, height, sensitivity to pollen, and so on, within a given race and group of healthy people. These ranges are quite normal, representing the complexity of, amongst other things, our genetic make-up. Based on all these factors, we would then be able to define a weight range for our subject. If he were above this range, then we could define him as being 'obese'. That is, there comes a point when the increasing weight of an individual crosses a borderline, so that he or she passes into another descriptive category. In the same way, 'individuality' turns into 'eccentricity' or even 'madness' if it becomes too marked.

Similarly, if the concentration of haemoglobin in the blood becomes low enough, it merits a diagnosis of anaemia; and if body temperature becomes raised sufficiently, it is said that we have a fever. Nearly always, the choice of a boundary is rather arbitrary; even the use of a range of values encompassing 95 per cent of the population is arbitrary. Why 95 per cent? Nevertheless, the use of such boundaries gives rise to standardized values that can become the basis for a diagnosis of 'obesity' or whatever.

With variables that show daily rhythms of high amplitude, another problem arises. Consider the case of trying to assess if the release into the bloodstream of growth hormone (the hormone that is mainly responsible for our increase in height during infancy and adolescence) is too low in an individual who is particularly short. As Fig. 3.4 shows, this hormone is normally released in bursts that are concentrated during sleep, particularly during the first hour or so. A sample of blood collected during the daytime, therefore, is unlikely to be useful, since a low concentration would be expected at this time. The correct time of sampling would be in the early hours of sleep. Conversely, if a patient is suspected of releasing too much of this hormone, then day-time sampling, when blood concentrations should be low, would be required.

Similarly, reference to the daily rhythm of the concentration of cortisol in the blood (see Fig. 3.4) indicates that, here also, the time of taking the blood sample is important. A suspected lack of secretion of the hormone should be assessed by taking samples in the morning on waking, and an excess, by taking them in the evening.

A rather unusual example of the importance of the correct time to sample blood is in the diagnosis of infection by certain threadworm parasites. Some species are found in the bloodstream of the patient in the daytime and 'hide' in the lungs at night. For other species, the reverse holds — the parasite being detectable in the blood only at night. Clearly, sampling at the appropriate times is essential in such cases.

In summary, if daily rhythmic changes in a variable are marked, then the time of diagnosis becomes important. In discussing symptoms with the patient, the doctor might concentrate on that time of day when the symptoms are worst, as in the case of rheumatoid arthritis, for example. Note that this advice would be difficult to implement reliably if the patient had recently flown from another time zone or had been working night shifts.

To put this into perspective, for many substances, the time of diagnosis is less important, because the range of daily changes is too small. For example, the daily rhythm in the concentration of haemoglobin in the blood is not an important consideration when diagnosing anaemia, since the decrease in this condition is far greater than occurs as a result of the daily rhythm of haemoglobin concentration.

## 14.4 Altered daily rhythms as a sign of disease

In some disorders, altered daily rhythms will contribute to some of the symptoms. Knowledge of this fact can act as an additional diagnostic aid, and the disappearance of abnormalities can help in the assessment of treatment. For example, in some cases of heart failure and renal disorder, particularly where fluid accumulation (oedema) has taken place, the rhythm of urine flow becomes inverted, with an increased, rather than decreased, flow at night. This might well lead to an increased frequency of waking at night to empty a full bladder, or even to bed-wetting. In either case, this information will be useful in building up a complete picture of the abnormality.

Whether or not we are waking at night with a need to pass urine can be seen as some form of self-diagnosis. In the same way, regular use of the bathroom scales not only enables us to see if we are putting on too much weight, but also if our diet is working. It has been argued that regular measurements of a daily rhythm by an individual will give a 'personalized' profile of that rhythm and that, with this information, small changes from the individual's norm can be picked up. The argument continues that such 'advance warning' can better enable countermeasures to be taken.

Apparatus for measuring one's own heart rate and blood pressure quickly and conveniently is now widely available. A high blood pressure throughout the daytime would be a matter for concern and would indicate the need to take medical advice. A high blood pressure at night is seen by some as an even greater cause for concern — though measuring this is more complicated. Waking up each hour is not useful, since it is the value when asleep that is required.

Earlier diagnosis and treatment of a disorder is advantageous, of course, but there is no clear evidence that altered daily rhythms reliably precede an illness (but see comments on asthma attacks, below). Also, there should be no doubt as to the huge amount of methodical measurement that would be required, and the reader is warned that converting large amounts of data into reliable estimates of daily rhythms is no easy matter.

Self-diagnosis has proved useful in the case of asthma sufferers, however. As already described, our airways are widest about noon and thereafter narrow gradually until they show the highest resistance to airflow during the night. If this narrowing becomes too marked, then an asthma attack, generally at night, might take place. It seems that, in some cases, the fall in the evening becomes more marked **in the days before an actual asthma attack.** That is, the asthmatic can predict the increasing likelihood of an attack, and take evasive action by increasing the amount of preventative drug that is being taken. Moreover, the patient can estimate if a treatment is continuing to be successful by making measurements of airway resistance. Knowledge of the daily rhythm of airway resistance (Fig. 14.1) indicates that the best times to make such measurements are in the evening or first thing on waking in the morning.

## 14.5 The timing of treatment

We have got used to the idea of taking pills 'three times per day with meals', just after having got up in the morning, or just before retiring at night. In many cases, this regimen is used mainly because it is a means of remembering when to take the pills regularly. However, the most appropriate timing of treatment might depend upon the severity of symptoms being experienced at a particular time.

### When the symptoms are worst

Consider the treatment of an asthma attack, indigestion, and a headache. Obviously, the asthmatic would inhale a substance that produces widening of the airways if an asthma attack were threatened or actually beginning; an individual suffering from indigestion would take a pill to combat this at the first sign of discomfort; and we would take an analgesic tablet as soon as a headache started.

### Preventative treatment

Many treatments are taken preventatively, the aim being to reduce the possibility of having the medical event occur in the first place. A simple (and pessimistic) example would be that of an individual who took a pill against indigestion before every meal, so preventing the appearance of the symptoms. A more pertinent example would be that of an insulin-dependent diabetic using a form of insulin that acts rapidly, and taking it at times closely linked to mealtimes. In these cases, it becomes important to ensure that a drug is present in the body when it will be most needed (just after a meal in both the cases just cited). There are further examples.

Attempts can be made to reduce the likelihood of asthmatic attacks by suitably-timed administration of a drug that suppresses the immune response (so mimicking the effect of cortisol). A suitable dosing schedule would ensure that the drug was present in the bloodstream at the right concentration during the evening and night, when attacks are most likely. For arthritic pain also, some of the drug (which also reduces the immune response) might be taken in the evening, so that the concentration throughout the night is maintained more effectively. Drugs and dosing regimens have been developed for both conditions to ensure that the drug is effective at the times it is required most.

To guard against the possibility of the formation of a blood clot after an operation, it is common to give an anticoagulant by continuous intravenous drip. As has already been described, the blood is more prone to clot in the hours before noon than in the evening. Therefore, if a constant rate of infusion of the drug were used (and this is the normal clinical practice), it might be found that this rate of infusion is too high in the evening (when the tendency for blood to clot is naturally low), producing an increased risk of haemorrhage; by contrast, the same rate of infusion might be too low in the morning (when the tendency for the blood to clot is higher), there still being a risk of clotting at this time. Note that this result implies that the best treatment in this case would be rhythmic — a point we shall return to later.

Attempts to reduce raised blood pressure originally focused on the high daytime values, and an antihypertensive drug was often given each morning. This also had the advantage that the patient was least likely to forget to take it then! Although these drugs reduced blood pressure in the daytime, their effects were wearing off by the next morning. The result was that the nocturnal blood pressure was not reduced, nor was the rise after waking up in the morning. However, when the rise in blood pressure was implicated in the increased frequency of cardiovascular problems at this time of day (see above), the treatment was altered. One solution was to divide the daily dose into two portions, to be taken in the morning and evening, so keeping blood pressure down at all times. A second solution has been to develop longer-acting drugs whose effects last a full 24 hours. There is increasing evidence that these recent treatments are more successful.

## Replacement therapy

Excessively short stature and delayed puberty are often due to the failure to secrete sufficient amounts of a particular hormone (growth hormone and sex hormones, respectively) into the bloodstream around the age of normal puberty. Originally, hormone replacement consisted of injecting the hormone into the patient and trying to raise its level throughout the 24 hours. This was often inconvenient to the patient, as well as being very expensive — and, anyway, it did not work! The bones or gonads (testes or ovaries) responded initially, but then became insensitive to the injected hormone. The daily rhythm of hormone release into the bloodstream had been ignored.

It is now known that the general rule is to mimic the daily rhythm of the hormone in blood as closely as possible. Initially, this seemed a complex process, because hormones are released into the blood as a series of pulses, so producing a jagged rather than a smooth 24-hour profile (see Fig. 3.4). However, it appears that a single injection each evening for both growth hormone (short stature) and gonadotrophic hormone (delayed puberty), so mimicking the time of greatest secretion of the hormone in normal subjects, is effective. This is also the reason why hormone replacement therapy for post-menopausal women (the hormones involved, oestrogens, are secreted mainly during sleep in pre-menopausal women) generally consists of taking a pill once a day on retiring.

## 14.6 Long-term treatment and chronopharmacology

In the examples of preventative treatment considered above, the treatment generally lasts for extended periods of time — sometimes for the rest of the life of the patient. In such circumstances, it is important to consider if the time of daily drug administration is optimal.

In some cases, it has been found that a drug does not always exert an identical effect upon the body, the effect depending partly upon the time of administration. These rhythmic changes include not only the therapeutic effects but also the undesirable side-effects of the drug. Clearly, we wish to choose a time of administration of the drug that will maximize the ratio of therapeutic effects to unwanted side-effects — a time when the drug will do most good and be least toxic and do least harm to the patient. There are two implications of this. First, it might be possible to choose a time of administration when the amount of drug that needs to be given to produce the required effect can be reduced, so reducing the cost of treatment too. Second, it might be possible to administer the drug at a time when it has less unpleasant side-effects, as a result of which it could be given at a greater dose and/or for a longer period of time.

This new study of the relationship between time and the effects of a drug is called chronopharmacology (Fig. 14.3).

As Fig. 14.3 indicates, the time-dependent effects of drug administration (the chronergy of the drug) depend upon an interaction between two factors:

- There are rhythmic changes in the sensitivity of the body to a fixed concentration of drug. This is a reflection of the daily rhythms in

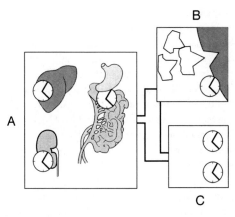

**Figure 14.3** Factors affecting the timing of drug treatment.
(A) chronokinetics; (B) chronesthesy; (C) chronergy (*see* text for details).

the number and properties of the sites on the cell surface which combine with the drug. This has been termed the chronesthesy of the body.

- When a drug is administered, there is a rise followed by a fall in its concentration in the blood. This represents the time course of the uptake of the drug, its distribution within the body, its breakdown by the liver, and its excretion by the kidneys. All of these factors show daily changes (as would be predicted from earlier chapters) and so alter the time course of the concentration of the drug in the blood and cells of the body. This phenomenon has been termed the chronopharmacokinetics of the drug.

These concepts of chronopharmacology can be illustrated by reference to cancer chemotherapy. In this branch of therapeutics, an understanding of chronopharmacology is beginning to pay dividends.

## Chronotherapy and cancer chemotherapy

The drugs that are used in cancer chemotherapy are designed to disrupt the process of cell division. Cancer cells divide repetitively and without showing a daily rhythm, and so they are similarly sensitive to the effects of such drugs at all times of the 24 hours. There are also some healthy tissues of the body where cell division is rapid — the bone marrow (which produces blood cells) and the cells of the gut, for example. For these tissues also, as with the cancerous tissue, high doses of the drugs used in chemotherapy can be toxic. Therefore, these healthy tissues must be protected as much as possible. In contrast to cancerous cells, these healthy tissues show daily rhythms of cell division, dividing rapidly at some times of the day and much less frequently at others. This difference in the daily rhythms of cell division between cancerous and healthy cells is one of the cornerstones of chemotherapy. Treatment regimens have been devised by which the drug is given at a time when cancerous cells are dividing rapidly (and so are susceptible to the drug) but the healthy cells of the tissue are much less active (and less susceptible).

The aim of a successful therapy, however, is not only to find a time of giving the drug that produces the best therapeutic effect (killing the cancerous rather than the healthy tissue), but also one which will not produce unacceptably large side-effects (killing or damaging the healthy issue, or producing nausea). If the side-effects can be reduced, then more drug can be given and for a longer period of time, both of which will increase the chance of success. When considering the undesirable side-effects of treatment, it has been found that, when the drug is excreted from the body via the kidneys, it tends to damage them. However, this side-effect can be decreased if the kidneys are producing a large volume of urine at the time, since this will dilute the drug that is being excreted. Therefore, the drug is given with large volumes of fluid at a time when the kidneys are most able to increase the flow of urine (see Chapter 6). As a result, more of the drug can be given to the patient without increasing the risk of kidney damage.

The field of timed administration of drugs is still in its infancy and is necessarily dominated by experiments upon animals. Nevertheless, the results of clinical trials on patients are now beginning to appear. They are very encouraging, with significant increases in the quality of life and survival time of the patients treated by these new, chronotherapeutic methods.

## The use of mini-infusion pumps

On several occasions — when considering the infusion of an anticoagulant, hormonal replacement therapy, and cancer chemotherapy — we have mentioned the desirability of varying the rate of infusion of a drug or hormone during the course of the 24 hours. Normally, this would demand too much time from medical personnel to be a feasible proposition. However, devices have been developed recently that might overcome these difficulties. They are small pumps that are implanted into the body and release a series of drugs. The drugs are stored in reservoirs and need to be replenished only every month or so. The device is powered by a small battery and controlled by a micro-computer which can be programmed from outside the body. This microcomputer controls the rate of release of the drugs, each drug hav-ing its own daily profile of release. The apparatus enables the doses of several drugs to be controlled, both during the course of a single day and, as treatment progresses, on a weekly basis. Such a device has not yet reached the stage where it is used routinely, but it is an exciting combination of miniaturization and computing technology on the one hand, and the application of a knowledge of daily rhythms to a diffi-cult medical problem on the other.

# Now try this: a scientific problem (III)

Drug X has been designed to reduce blood pressure in patients with high blood pressure. Figure 14.4 shows a healthy subject's mean arter-ial blood pressure during the course of a normal 24-hour period; the mean arterial blood pressure of our (hypothetical) patient living a nor-mal life before treatment; and the mean arterial blood pressure of this patient during the 24 hours following treatment with drug X at 10:00.

Does drug X control blood pressure? Our comments on this problem follow.

## Comments

In many senses, X does control blood pressure. It lowers blood pressure to fairly normal values from about 14:00 (when the drug begins to take effect) until about 07:00 the following morning. Thus, it appears to be effective during the afternoon, evening, and most of the night. This would include times when our patient was awake and active as well as when he/she was asleep. By contrast, from 08:00 to 11:00, the blood pressure begins to rise further above the values for a healthy individ-ual, and to return towards the 'normal' (high) values for the patient. That is, drug X did not work after about 07:00.

The effect of this is that there is a very marked rise in blood pres-sure between 08:00 and 10:00, and this time coincides with when some fatal heart disorders are most common. This might be a cause for con-siderable concern. Possibly, the drug is no longer working by that stage (more than 21 hours after its administration) or its mechanism of

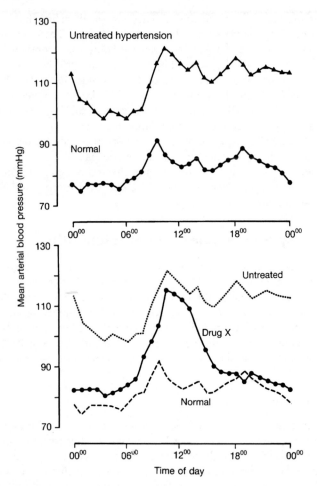

**Figure 14.4** Top: mean arterial blood pressure in a healthy subject and in a patient with high blood pressure (hypertension). Bottom: hypothetical effects on blood pressure of administering drug X at 10:00.

action is such that it cannot reduce the morning rise. Either taking the whole dose of X at 22:00, or taking half the dose at 10:00 and half at 22:00, are just two of the ways of deciding between such possibilities.

If the problem is that the drug only works for about 21 hours after administration, then it is possible that this can be remedied quite easily — either by taking the drug in the evening, or by taking half the dose in the morning on waking and half in the evening before retiring, or by using a form of the drug that is released into the bloodstream more slowly. If the rapid rise between 08:00 and 10:00 cannot be prevented by any of these alternative dosing regimens (and this would be the case if, for example, the drug did not prevent the effects due to the rise of sympathetic nervous system activity in the hours immediately after waking), then severe doubts would be cast on the value of X. In this case, some slight change to the chemical structure of X might produce a better drug — but it is a time-consuming and costly business to develop new drugs.

This example, like that of the chronotherapy of cancer, illustrates that knowledge of the role that time can play in the body in disease and clinical disorder can lead to the development of better therapies that improve the quality of life and life expectancy of the patients.

# Index

Lightning Source UK Ltd.
Milton Keynes UK
UKOW051327051212

203201UK00005B/35/P